ENGINEERING THERMODYNAMICS

ENGINEERING THERMODYNAMICS

An Introduction

M. S. Kassim, Ph.D.
D. S. Khudhur, Ph.D.
H. M. Hussain, Ph.D.
L. J. Habeeb, Ph.D.

MERCURY LEARNING AND INFORMATION
Dulles, Virginia
Boston, Massachusetts
New Delhi

Publisher: David Pallai
MERCURY LEARNING AND INFORMATION
22841 Quicksilver Drive
Dulles, VA 20166
info@merclearning.com
www.merclearning.com
1-800-232-0223

M. Kassim, D. Khudhur, H. Hussain, L. Habeeb. *Engineering Thermodynamics*.
ISBN: 978-1-68392-859-1

The publisher recognizes and respects all marks used by companies, manufacturers, and developers as a means to distinguish their products. All brand names and product names mentioned in this book are trademarks or service marks of their respective companies. Any omission or misuse (of any kind) of service marks or trademarks, etc. is not an attempt to infringe on the property of others.

Library of Congress Control Number: 2022932252

222324321   Printed on acid-free paper in the United States of America.

Our titles are available for adoption, license, or bulk purchase by institutions, corporations, etc. For additional information, please contact the Customer Service Dept. at 800-232-0223(toll free).

All of our titles are available in digital format at *academiccourseware.com* and other digital vendors. The sole obligation of MERCURY LEARNING AND INFORMATION to the purchaser is to replace the book, based on defective materials or faulty workmanship, but not based on the operation or functionality of the product.

CONTENTS

CHAPTER 1

INTRODUCTION

1.1   DEFINITION AND SCOPE OF THERMODYNAMICS

Thermodynamics is a basic science that deals with the conversion of heat into mechanical work. Nowadays, the common concept of thermodynamics is that it is concerned with the understanding of changes in behavior and properties of matter affected by temperature change.

This concept is also known as energy science, establishing relations among work, heat, and features of the systems, which are in equilibrium. It depicts the changes in the physical systems through an interaction of energy. Energy is neither formed nor annihilated and total energy remains constant.

This concept deals with real physical systems that might be solid, fluid, mixtures, or even empty space with electromagnetic waves. The system should be some type of vessel where no chemical reaction takes place.

This basic science studies changes in the properties or behavior of the fluid cooled down or heated. The fluid may be a gas (such as air), vapor (such as water vapor), liquid, or a mixture of the substances provided they do not react chemically with each other.

It also studies the relationship between changes of fluid properties and quantities of work and heat causing this change.

Thermodynamics is said to be based on four basic principles or laws found through experience and not through mathematical derivation.

These laws are as follows:

1. **Zeroth law** deals with thermal equilibrium under which temperature is defined. It is called zeroth law because it was drafted after the first law.

2. **First law** is the law of the conservation of energy, that is, energy can neither be created nor destroyed.

3. **Second law** determines the direction of the process or the direction of the energy flow and the percentage of energy conversion.

4. **Third law** determines the entropy and shows the impossibility to reach absolute zero entropy. Entropy has zero value at absolute zero temperature.

Engineers use this science to design thermal engines such as power plants, reciprocating engines, jets and missiles, gas and steam turbines, steam boilers, compressors, air conditioners, and other machines and plants. It is very necessary for engineers to understand the concepts and laws of thermodynamics.

The conversion of heat into mechanical work has been known since the 18th century. In the middle of the 19th century, a scientist named James Prescott Joule created the relationship between mechanical work and thermal energy. Many scientists have contributed to the development of thermodynamics such as Carnot, Kelvin, Clausius, and others. The present-day thermodynamics deals with all thermal machines and air conditioners, which are consumers of energy.

1.2   HEAT AND WORK

Heat is the energy transferred between the system and its surroundings due to the difference of temperature only. There are three methods for the transfer of heat: convection, radiation, and conduction.

Heat is also considered as a method of altering the energy content of the system by the effective variance of temperature. Another method for altering the energy of a system is called work. We can define work as electric work (in an electric motor), mechanical work (in a piston-cylinder and engine), and chemical work.

1.2.1 Types of Mechanical Work

In mechanics, when a force F travels through a distance dx, work is being done. The examples of mechanical work are as follows:

1. the elasticity of a spring,

2. moving system boundary or displacement or flow of energy,

3. surface tension, and

4. shaft rotation on shaft energy.

The concept of thermodynamics was developed to understand the limitation of the efficiency of an engine. Modern power plants work on complicated cycles when compared with the operation of original steam engines, which had only a small percentage of effectiveness in the conversion of heat into work, whereas new utility steam power plants or gas turbine plants have efficiency higher than 60%.

1.2.2 Forms of Energy

There are many forms of energy such as potential, kinetic, thermal, magnetic, chemical, mechanical, nuclear, electrical, etc.

If E is the total energy of a system, the specific energy indicated by e is stated as energy per unit mass:

$$e = \frac{E}{m}(\text{kJ} / \text{kg})$$

Energy cannot be destroyed or created. It can only be transferred or changed from one form to another, that is, energy is conserved. Electrical energy, thermal energy, or mechanical energy can be transformed from one form to another. Heat engines convert thermal energy into mechanical energy.

1.2.3 Displacement Work

Work is done when the point of application of a force moves in the direction of force:

$$W = F \times x$$

where W represents work, F represents force, and x represents displacement.

If the force changes with displacement, then

$$W = \int_{1}^{2} F(x)\,dx \qquad (1.1)$$

or

$$_1W_2 = \int_{1}^{2} F\,dx \qquad (1.2)$$

Figure 1.1 shows a piston-cylinder arrangement. In position 1, a gas at pressure p is confined inside a cylinder. "A" represents the cross-sectional area of the cylinder. Distance X represents the top of the piston in the cylinder. Some weights may be placed on the piston top to balance gas pressure in a stable state.

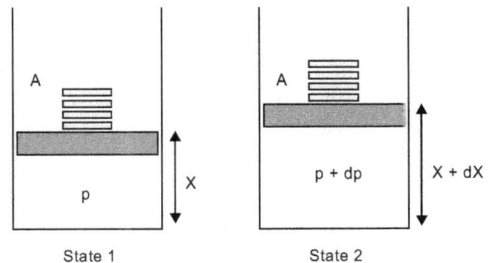

FIGURE 1.1 Piston cylinder-arrangement.

In state 2, the piston moves to a new height $X + dX$ with $dp + p$ as a new pressure in the cylinder to balance a new force of weight.

Obviously work is done as a force acts through distance dX. The amount of work done is given as follows:

$$\delta W = F\, dx \tag{1.3}$$

But, F varies during the process.

For state 1, force $F = p \times A$

For state 2, force $F = (p + dp) \times A$

The average value of force is given as follows:

$$F = \frac{1}{2}\left[pA + (p + dp)A\right] = pA + \frac{dp}{2}A.$$

$$\therefore \qquad \delta W = \left(pA + \frac{dp}{2}A\right)dX = pA\, dX + \frac{A}{2}dp\, dX$$

Neglect $dp\, dX$ as very small

$$\therefore \qquad \delta W = p \times A \times dX \tag{1.4}$$

Now, $dV = A \times dX$, where dV is the difference in volume, and the significant equation will be given as:

$$\delta W = p \times dV \tag{1.5}$$

By integration,

$$_1W_2 = \int_1^2 p\, dV \tag{1.6}$$

$_1W_2$ represents the area under a curve in a $p - V$ graph. Figure 1.2 shows two processes A and B joining the same points 1 and 2.

The work $_1W_2$ depends on the selected path, and not the endpoints. The work is the area on $p - V$ diagram beneath the process line.

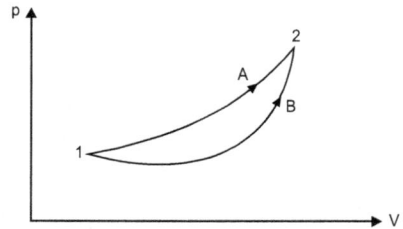

FIGURE 1.2 $p - V$ graph of work for two diverse procedures joining the similar state.

1.2.4 Polytropic Processes

The polytropic process is explained by the following equation:

$$p \times V^n = C = \text{constant}$$

n is identified as the polytropic index as presented in Figure 1.3.

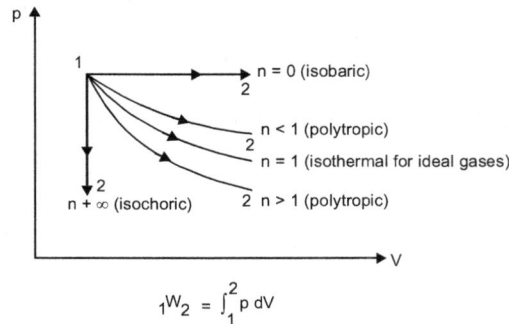

FIGURE 1.3 $p - V$ diagram for various polytropic processes.

Work is done by a system if the sole effect on the surrounding (i e., the things exterior to the system) could be reduced to the rising of a weight.

Sign convention

- Work done by the system on the surrounding is positive.
- Work done by the surrounding system is negative.

1.2.5 Other Forms of Work

In addition to pressure forces, there are other types of forces. These forces also can do the work:

- A wire is elongated by a force of tension τ through length dL. The work done is given as follows:

$$\delta W = -\tau dL \tag{1.7}$$

- Surface tension δ. The differential work is given as follows:

$$\delta W = -\delta \, dA \tag{1.8}$$

- Electrical work in a system having ε as potential and electrical charge, q of an atom, and the distance as x, will be

$$\delta W = -q\varepsilon \, dx \tag{1.9}$$

In general,

$$\delta W = -p \, dV = \tau \, dL = \delta \, dA = q \, \varepsilon \, dx \tag{1.10}$$

The work done by gas expanding in a vacuum is given as:

$$_1W_2 \neq \int_1^2 p \, dV$$

Because it is a non-steady process.

1.2.6 Work Summary

TABLE 1.1 Work summary for an ideal gas (unit mass)

Process	Index N	Heat added	$\int_1^2 p \, dv$	p, v, t relations	Specific heat, c
Constant pressure	$n = 0$	$c_p(T_2 - T_1)$	$p(v_2 - v_1)$	$\dfrac{T_2}{T_1} = \dfrac{v_2}{v_1}$	c_p
Constant volume	$n = \infty$	$c_v(T_2 - T_1)$	0	$\dfrac{T_1}{T_2} = \dfrac{P_1}{P_2}$	c_v
Constant Temperature	$n = 1$	$p_1 v_1 \log e \dfrac{v_2}{v_1}$	$p_1 v_1 \log e \dfrac{v_2}{v_1}$	$p_1 v_1 = p_2 v_2$	∞
Reversible Adiabatic	$n = \gamma$	0	$\dfrac{p_1 v_1 - p_2 v_2}{\gamma - 1}$	$p_1 v_1{}^\gamma = p_2 v_2{}^\gamma$ $\dfrac{T_2}{T_1} = \left(\dfrac{v_1}{v_2}\right)^{\gamma-1}$ $= \left(\dfrac{p_2}{p_1}\right)^{\frac{\gamma-1}{\gamma}}$	0
Polytropic	$n = n$	$e_n(T_2 - T_1)$ $= c_v\left(\dfrac{\gamma - n}{1 - n}\right) \times (T_2 - T_1)$ $= \dfrac{\gamma - n}{\gamma - 1} \times \text{work done}$ (non-flow)	$\dfrac{p_1 v_1 - p_2 v_2}{n - 1}$	$p_1 v_1{}^n = p_2 v_2{}^n$ $\dfrac{T_2}{T_1} = \left(\dfrac{v_1}{v_2}\right)^{n-1}$ $= \left(\dfrac{p_2}{p_1}\right)^{\frac{n-1}{n}}$	$c_v\left(\dfrac{\gamma - n}{1 - n}\right)$

EXAMPLE 1.1

In a piston-cylinder arrangement, the gas is initially at 150 kPa and occupies a volume of 0.03 m³. The gas is heated so that the volume of gas increases to 0.1 m³. Calculate the work done by the gas if the volume is inversely proportional to the pressure.

Solution:

$$p_1 = 150 \text{ kpa}$$

$$V_1 = 0.03 \text{ m}^3$$

$$V_2 = 0.1 \text{ m}^3$$

$$V \propto \frac{1}{p}$$

$$pV = C = p_1 V_1 = p_2 V_2$$

$$p = \frac{C}{V}$$

$$\therefore W_{1-2} = \int p \, dV = C \int_{V_1}^{V_2} \frac{dV}{V} = C \, ln \frac{V_2}{V_1} = p_1 V_1 \, ln \, p_1 V_1$$

$$= 150 \times 0.03 \, ln \frac{0.1}{0.03} = 5.41 \text{ kJ} \quad \textbf{Ans.}$$

EXAMPLE 1.2

Compute the work which a pump shall have to do upon water in an hour to just force the water into a tank (closed) having a pressure of 1.0 MP at the rate of 1.5 m³/min horizontally from an open well.

Solution:

$$p = 1 \text{MPa} = 1 \times 10^3 \text{ kPa}$$

$$V = 1.5 \text{m}^3 / \text{min} = 1.5 \times 60 = 90 \text{m}^3 / \text{hr}$$

$$W = pV = 1 \times 10^3 \times 90$$

$$= 90 \times 10^3 \text{ kJ} \quad \textbf{Ans.}$$

EXAMPLE 1.3

The flow energy of 0.124 m³/min of a fluid crossing a boundary of a system is 18 kW. Find the pressure at this point.

Solution:

$$\text{Flow energy}, FE = 18\,\text{kW} = 18 \times 10^3\,\text{W}$$

$$V = 0.124\,\text{m}^3 / \text{min} = \frac{0.124}{60}\,\text{m}^3 / \text{s}$$

Now

$$FE = pV$$

$$\therefore \quad p = \frac{FE}{V} = \frac{18 \times 10^3}{0.124 / 60}$$

$$= 8.7 \times 10^6\,\text{N/m}^2$$

$$= 87\,\text{bar} \quad \textbf{Ans.}$$

EXAMPLE 1.4

Determine the size of a spherical balloon filled with hydrogen at 30°C and atmospheric pressure for lifting 400 kg payload. Atmospheric air is at a temperature of 27°C and the barometer reading is 75 cm of Hg.

Solution:

$$\text{Assume of balloon} = V\text{m}^3$$

$$\text{Pressure}\, p_1 = \rho\, g\, h$$

$$= 13.6 \times 10^3 \times 9.81 \times \frac{75}{100} \times 10^{-5} = 1\,\text{bar}$$

Gas constant for hydrogen:

$$R = \frac{R}{M} = \frac{8314}{2} = 4157\,\text{J/kg K}$$

$$\text{Temperature},\, T = 27 + 273 = 300\,\text{K}$$

Apply equation of state:

$$p_1 V_1 = m_1 R T_1$$

Mass of hydrogen filling the balloon:

$$m_1 = \frac{p_1 V_1}{R T_1} = \frac{1 \times 10^5 \times V}{4157 \times (273 + 30)} = 0.079392\,V\,\text{kg}$$

Mass of atmospheric air displaced,

$$m_2 = \frac{p_1 V_1}{RT_1} = \frac{1 \times 10^5 \times V}{287 \times 300} = 1.16144 \text{ V kg}$$

$$\text{Pay load} = m_2 - m_1$$

$$= 1.6144 - 0.079392 \text{ V}$$

$$= 400 \text{ kg}$$

$$\therefore \quad V = 369.669 \text{ m}^3 \textbf{ Ans.}$$

EXAMPLE 1.5

Three kg of air kept at 100 kPa and 300 K is compressed polytropically to 1500 kPa and 500 K.

Calculate:

    (*i*) **index,**
    (*ii*) **final volume,**
    (*iii*) **work done, and**
    (*iv*) **heat exchanged.**

Solution: (*i*) **Index**

$$p_1 = 100 \text{ kPa} = 100 \times 10^3 \text{ N/m}^2$$

$$T_1 = 300 \text{ K}$$

$$p_2 = 1500 \text{ kPa} = 1500 \times 10^3 \text{ N/m}^2$$

$$T_2 = 500 \text{ K}$$

$$p_1 V_1^n = p_2 V_2^n$$

But

$$\frac{p_1 V_1}{T_1} = \frac{p_2 V_2}{T_2}$$

$$\frac{T_2}{T_1} = \left(\frac{p_2}{p_1}\right)^{\frac{n-1}{n}}$$

$$\frac{n-1}{n} = \frac{ln \dfrac{T_2}{T_1}}{ln \dfrac{p_2}{p_1}} = \frac{ln \dfrac{500}{300}}{ln \dfrac{1500}{100}} = 0.1886$$

$$\therefore n = 1.23 \textbf{ Ans.}$$

(ii) Final volume

$$p_1 V_1 = mRT_1$$

Take $R = 287$ J/kg K for air

$$V_1 = \frac{mRT_1}{p_1} = \frac{3 \times 287 \times 300}{100 \times 10^3} = 2.583\,\text{m}^3$$

$$p_2 V_2 = mRT_2$$

$$V_2 = \frac{mRT_2}{p_2} = \frac{3 \times 287 \times 500}{1500 \times 10^3} = 0.287\,\text{m}^3 \;\textbf{Ans.}$$

(iii) Work done

$$W_{12} = \frac{p_1 V_1 - p_2 V_2}{n-1} = \frac{mR(T_1 - T_2)}{n-1}$$

$$= \frac{3 \times 287\,(300 - 500)}{1.23 - 1}$$

$$= -748696\,\text{J} = -748.7\;\text{kJ}$$

Work is done in the air during compression.

(iv) Heat exchanged

$$Q_{1-2} = \frac{\gamma - n}{\gamma - 1} W_{1-2}$$

$$= \frac{1.4 - 1.23}{1.4 - 1}(-748.7)$$

$$= -318.2\;\text{kJ}\quad\textbf{Ans.}$$

Heat is rejected by the system.

EXAMPLE 1.6

Calculate the work done in a piston-cylinder arrangement during an expansion process, where the pressure is given by the equation:

$$p = (V^2 + 6V)\;\text{bar}$$

The volume changes from 1 m³ to 4 m³ during expansion.

Solution:

$$V_1 = 1\;\text{m}^3$$

$$V_2 = 4\;\text{m}^3$$

$$W_{1-2} = \int_1^2 p \, dV$$

$$= \int_{V_1}^{V_2} (V^2 + 6V) \times 10^5 \, dV$$

$$= 10^5 \left| \frac{V^2}{3} + \frac{6V^2}{2} \right|_1^4$$

$$= 10^5 \left[\frac{(4^3 - 1^3)}{3} + 15 \times 3 \right] = 66 \times 10^5 \text{ J}$$

$$= 6.6 \text{ MJ} \quad \textbf{Ans.}$$

EXAMPLE 1.7

The cylinder of a compressor contains 1 kg of NH_3. The initial pressure of NH_3 is 2 MPa, and the temperature is 180°C. Cooling the gas to 40°C, then to 20°C, at this state the NH_3 quality is 50%. Determine the overall work for the process, which is a direct function of the difference of V against p.

Solution:

Stage 1: (P, T) from data in Table B.2.2

The value of $v_1 = 0.1057100$ m³/kg

Stage 2: (x, T) from data in table B.2.1 saturated vapor.

$$v_2 = 0.0831300 \text{ m}^3/\text{kg}$$
$$p_2 = 1555.00 \text{ kPa}$$

Stage 3: (x, T) $\quad p_3 = 857.00 \text{ kPa}$

$$v_3 = \frac{(0.1492200 + 0.00163800)}{2} = 0.0754300 \text{ m}^3/\text{kg}$$

The area under the curve of the graph $p - V$ gives total work:

$$_1W_3 = \int_1^3 p \, dV \approx m \times \left(\frac{(P_2 + P_1)}{2} \right) \times \frac{(-v_1 + v_2) + m(P_3 + P_2)(-v_2 + v_3)}{2}$$

$$_1W_3 = \frac{(2000.000 + 1555.000)}{2}$$

$$\frac{1(0.0831300 - 0.1057100) + (1555.00 + 857.00)}{2} 1(0.0754300 - 0.0831300)$$

$$= -49.400 \text{ kJ}$$

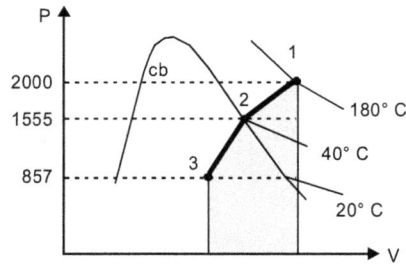

FIGURE 1.4

1.2.7 Heat

Heat is a form of energy, which is transferred across the boundary of a system at a definite temperature to another system at a different temperature due to the difference in temperature between the two.

There are three modes of transfer of heat:

1. **Conduction** is also called heat spread due to some factors. Bacon is cooked by conduction. This is explained by Fourier's law.

   For one dimensional heat conduction:

   $$q = -k\frac{\Delta T}{\Delta x} \tag{1.11}$$

   $$Q = qA$$

   $$Q = -kA\frac{dT}{dx} = -kA\frac{T_{hot} - T_{cold}}{dx} \text{(J/S or W)} \tag{1.12}$$

2. **Convection:** The conduction is enhanced by the flow of fluid as seen in Figure 1.5 Convective effects are explained by Newton's law of cooling:

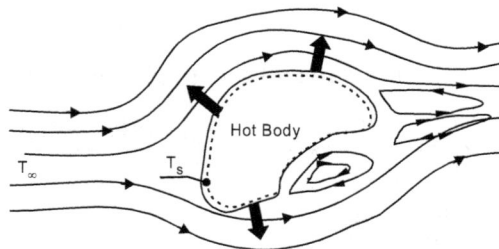

FIGURE 1.5 Convective heat transfer from the hot body.

$$q = h(T_{hot} - T_{cold}) \tag{1.13}$$

$$Q = qA = hA(T_{hot} - T_{cold})(W/m^2/K) \tag{1.14}$$

where h is a constant.

3. **Thermal radiation:** The sun heats the earth by radiation. This is due to the remote effects of energy diffusion. The rate of heat transfer by radiation is given by:

$$q = \sigma\left(T_{hot}^4 - T_{cold}^4\right) \tag{1.15}$$

and $$Q = qA = \sigma A\left(T_{hot}^4 - T_{cold}^4\right) \tag{1.16}$$

where $$\sigma = 0.0567 \times 10^{-6}\left(W.K^4/m^2\right)$$

σ is called Stefan–Boltzmann constant.

The sign convection of heat transfer is as follows:

- Heat entering into the system is +ve.
- Heat leaving from the system is –ve.

EXAMPLE 1.8

A piston/cylinder engine contains water vapor of quality 0.7. The pressure is 200.00 kPa at an initial volume of 0.1 m³. The temperature of the system is raised to 200.00°C. Estimate the heat and the work transfer by the system.

Solution: Control volume contains the water vapor of 0.7 dryness fraction.

Equation of continuity: $m = m_1 = m_2$;

Equation of energy 5.11: $\left({}_1W_2 - {}_1Q_2\right) = -m(u_1 - u_2)$

When p is constant $_1W_2 = \int P \times dV = -P \times m(v_1 - v_2)$

Stage 1: Refer to steam Table B.1.2

Saturation temperature at
$$200.00kPa = 120.2300°C = T_1, v_{fg} = 0.88467 \text{ and } v_f = 0.001061$$

$$x \times v_{fg} + v_f = (0.7 \times 0.88467) + 0.001061$$
$$= 0.6203300\,\text{m}^3/\text{kg} = v_1$$
$$h_1 = x \times h_{fg} + h_f$$
$$= (0.7 \times 2201.96) + 504.68 = 2046.052 \text{ kJ/kg}$$

From the first conditions, the mass, m, is given by:

$$m = \frac{1}{v_1/V_1} = \frac{1}{0.6203300/0.1} = 0.161204\,\text{kg}$$

From the steam table, we get

$$v^2 = 1.0803400\,\text{m}^3/\text{kg at } p_2 = 200.00 \text{ kPa}$$
$$T_2 = 200.00°\text{C}$$
$$h_2 = 2870.4600 \text{ kJ/kg (Steam table)}$$
$$V_2 = v_2 \times m$$
$$= 1.0803400 \times 0.161204\,\text{m}^3/\text{kg} = 0.17415\,\text{m}^3$$

From the equation of energy, the work is:

$$_1W_2 + U_2 - U_1 = (u_2 + pv_2 - u_1 - pv_1)m$$
$$= {}_1Q_2 - m(h_1 - h_2)$$

The value of heat is given by:

$$_1Q_2 = -0.161204 \text{ kg} \times (2046.05 - 2870.460) = 132.9 \text{ kJ} \quad \textbf{Ans.}$$

1.3 THERMODYNAMIC SYSTEMS

A system in thermodynamics is a quantity of matter for the study and analysis of a problem. The system needs more surveys. Systems can be a simple cylinder containing gas or as complicated as a nuclear power plant. The whole region outside the system is called the surrounding. Systems are separated from the surroundings by real or imaginary limits known as boundaries as shown in Figure 1.6.

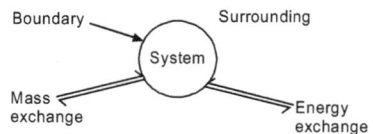

FIGURE 1.6 Thermodynamic system.

The boundary may be fixed or movable and real or imaginary.

The boundary may be in motion or at rest.

- Heat could be exchanged between the system and the surrounding.
- Work could be exchanged between the system and the surrounding.

Thermodynamic systems could be classified according to constant mass or a constant volume of the system.

1.3.1 Closed System

It is a control mass system that contains a fixed amount of matter. No mass can leave or enter a closed system border as seen in Figure 1.7.

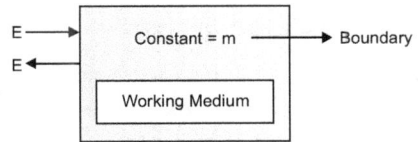

FIGURE 1.7 Closed system (control mass).

Mass cannot cross the borders of a closed system. The energy can be exchanged at the boundary.

Energy can be either heat or work. The volume of a closed system can change and may not be constant. As a special case, an isolated system is defined as a system that is not affected by its surroundings and there is no mass or energy exchange.

1.3.2 Open System

In an open system, both energy (heat and work) and mass can cross the boundary, for example, nozzle, turbine, or compressor. These equipments are studied as the control volume system, where energy and mass can pass its border (Figure 1.8) but volume remains constant or fixed.

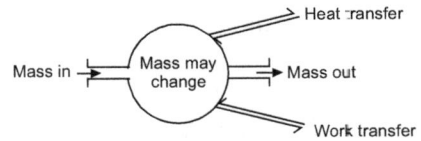

FIGURE 1.8 Open system.

Figure 1.9 shows energy interaction and one or many of the system properties may change. The system might reach a new equilibrium state.

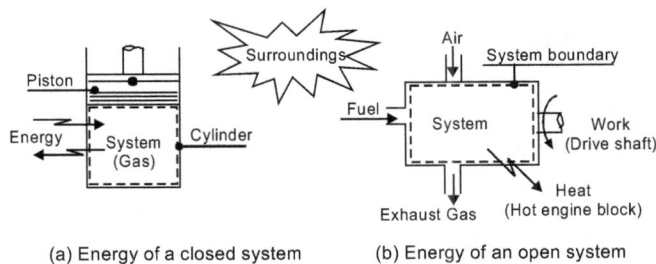

(a) Energy of a closed system   (b) Energy of an open system

FIGURE 1.9 System energy.

A large number of engineering devices are open systems where the flow of matter takes place into the system and out of the system. These are designed as control volume systems. For example, a car radiator, a turbine, a compressor, and a water heater.

(i) **Control Volume system:** It has constant volume with mass passing through its border.

(ii) **Control Surface:** The border of the control volume system is called Control Surface.

1.3.3 Isolated System

Figure 1.10 shows an isolated system. In this system, there is no exchange or transfer of mass and energy between the system and the surrounding.

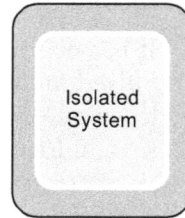

Isolated System

1.3.4 Adiabatic System

FIGURE 1.10 Isolated system.

It is a system that is thermally insulated at its boundary. It has a fixed thermal reservoir. If there is no work exchange, it can be an insulated system.

1.3.5 Similar System (Homogeneous System)

A phase can be defined as a matter with a similar physical structure and chemical composition. The system with single-phase matter is defined as a similar or homogeneous system. For example, a mixture of nitric acid with water, heptane with octane, and vapor of water with air.

1.3.6 Dissimilar System (Heterogeneous System)

A system that contains matter in two or more phases is called a heterogeneous system. Example: mixture of steam with water or water with ice.

1.4 MACRO AND MICRO APPROACH TO STUDY THE SYSTEM MATTER

A thermodynamic system may be a rod of steel or a vessel containing gas at some pressure. Most of the systems in engineering are gas containers. As per Avogadro's Hypothesis, an equal volume of all gases under identical conditions of temperature and pressure contain same number of molecules. The number of molecules may be 10^{23} moving randomly in all directions causing a collision of the order of 10^9 per unit time. They are separated from each other by a distance of the order of main-free path. The matter is neither continuous nor homogeneous from a microscopic point of view.

The thermodynamic properties are found out by statistically averaging the behavior of individual molecules. The results obtained are very accurate. For research and study of the law of thermodynamics deeply, statistical thermo-dynamics is used.

The microscopic approach is rather complex, cumbersome and time-consuming. However, engineering systems working at high pressure contain the large number of molecules in comparison to small volumes. The linear dimensions of the system are quite large as compared with inter-molecular distance, which can be neglected. It is assumed that the system contains the continuous distribution of matter with no voids or cavities. The system is regarded as a continuum. The macroscopic approach is based on the continuum hypothesis. All intensive properties such as pressure, temperature and density are regarded as definite values. The results shown are sufficiently accurate. The study is simple and quick.

The variance between microscopic and macroscopic approaches is given in Table 1.2.

TABLE 1.2

Macroscopic approach	Microscopic approach
A certain amount of material is measured without taking into consideration the behavior of the individual molecule. This approach is concerned with gross or overall behavior of system material.	The system is made of a large number of separate particles called molecules. These molecules have different velocities and energies. The values of these energies are constantly changing with time.
This is recognized as conventional thermodynamic or classical thermodynamics.	The approach is related to statistical thermodynamics.

1.5   SYSTEM STABILITY

In mechanics, a body is stable if it is in equilibrium. The principle of equilibrium states, "A stationary body subjected to a system of forces will be in equilibrium if the resultant of the forces and resultant moment are zero."

The properties of the system are measured and processes are studied when the systems are stable or in thermodynamic equilibrium. A system having the same values of properties (pressure, temperature, velocity, elevation, composition, etc.) throughout is in thermodynamic equilibrium. The system should be simultaneously in the state of mechanical, thermal, electrical, and chemical equilibrium.

1. **Thermal equilibrium** denotes uniformity of temperature or absence of temperature gradient or heat flow.

2. **Mechanical equilibrium** denotes uniformity of pressure or absence of unbalanced forces.

3. **Electrical equilibrium** denotes the absence of electrical potential and the absence of current flow.

4. **Chemical equilibrium** denotes the absence of phase change or chemical reaction and there is no mass diffusion.

1.6 PROPERTY RELATIONS

The functional relationship among the thermodynamic properties such as pressure, volume and temperature is known as the equation of state or characteristic gas equation.

The equation can be developed by combining Boyle's Law and Charles's Law for a perfect gas:

$$\left.\begin{array}{l} pv = RT \\ pV = mRT \\ p\bar{v} = MRT \\ pV = n(MR)T \end{array}\right\} \tag{1.17}$$

The above equations are called characteristic gas equations:

where

$$P = \text{pressure of gas} (\text{N/m}^2)$$

$$V = \text{volume of gas} (\text{m}^3)$$

$$v = \text{specific volume}$$

$$= \frac{V}{m} (\text{m}^3/\text{kg})$$

$$\bar{v} = \text{molar volume} (\text{m}^3/\text{mol})$$

$$T = \text{absolute temperature of gas (K)}$$

$$R = \text{characteristic gas constant} (\text{N-m/kg-K})$$

$$M = \text{Molar mass of gas}$$

$$n = \text{number of moles of the gas}$$

1.6.1 Universal Gas Constant

The universal gas constant or molar constant $(\overline{R})$ of a gas is the product of the characteristic gas constant (R) and the molecular mass (M) of the gas.

$$\overline{R} = MR$$

The value of $(\overline{R})$ is the same for all gases and can be derived from Avogadro's law where:

$$p = 760 \text{ mm Hg } (1.013 \times 10^5 \text{N/m}^2)$$

$$T = 273.15 \text{ K}$$

$$\overline{v} = 22.4 \text{ m}^3/\text{kg}$$

$$\overline{R} = \frac{1.013 \times 10^5 \times 22.4}{273.15}$$

$$= 8.3143 \text{ (kJ/kg mol K)}.$$

EXAMPLE 1.9

A tank of 0.35 m^3 capacity contains H$_2$S gas at 300 K. When 2.5 kg of gas is withdrawn, the temperature and pressure in the tank become 288 K and 10.5 bar, respectively. Calculate the mass of the gas initially present in the tank. Also, determine the initial pressure of the gas.

Solution:

$$V_1 = V_2 = 0.35 \text{ m}^3 \text{ (constant)}$$

$$T_1 = 300 \text{ K}$$

$$T_2 = 288 \text{ K}$$

$$p_2 = 10.5 \text{ bar} = 10.5 \times 10^5 \text{ N/m}^2$$

The molecular mass of H$_2$S gas is given as:

$$M = 2 + 32 = 34$$

$$R = \frac{\overline{R}}{M} = \frac{8314}{34} = 244.53 \text{ J/kg K}$$

Applying equation of state for final condition

$$p_2 V_2 = m_2 R T_2$$

$$\therefore \quad m_2 = \frac{p_2 V_2}{R_2 T_2} = \frac{10.5 \times 10^5 \times 0.35}{244.53 \times 288} = 5.2 \text{ kg}$$

$$\therefore \quad m_1 = 5.2 + 2.5 = 7.5 \text{ kg} \quad \textbf{Ans.}$$

Applying equation of state for the initial condition

$$p_1 V_1 = m_1 R T_1$$

$$\therefore \quad p_1 = \frac{m_1 R T_1}{V_1} = \frac{7.7 \times 244.53 \times 300}{0.35} = \mathbf{16.14 \ bar \ Ans.}$$

REVIEW QUESTIONS

1. Define thermodynamics. Discuss its scope. List the laws of thermodynamics.

2. Differentiate between work and heat transfer.

3. Explain in detail about displacement work. What are other forms of work?

4. What is a thermodynamic system? List various types of systems.

5. Differentiate between:
   a. Open and closed systems.
   b. Insulated and adiabatic systems.
   c. Similar and dissimilar systems.

6. Explain the following:
   a. Continuum
   b. Phase
   c. Elastic work
   d. Polytropic index

7. Compare macro and micro approaches to the study of system matter.

8. Write notes on:
   a. Thermodynamic equilibrium
   b. Universal gas constant

9. Establish a relationship for an ideal gas. What is the significance of the equation of state?

NUMERICAL EXERCISES

1. Estimate the heat required to raise the temperature of fluid from 300 K to 400 K if specific heat is given by

$$C = (0.2 + 0.002\,T)\ \text{kJ/kg.K}$$

   The mass of fluid is 2 kg. What will be the mean specific heat?
   (180 kJ, 0.9 kJ/kg K)

2. One mole of an ideal gas at 0.1 MPa and 300 K is heated at constant pressure till the volume is doubled and then it is allowed to expand at constant temperature until the volume is doubled again. Calculate the work done by the gas. **(6.15 MJ)**

3. Calculate the change of enthalpy as 1 kg of oxygen is heated from 500 K to 2,000 K. The value of specific heat constant pressure is given as

$$C_p = 11.515 - \frac{172}{\sqrt{T}} - \frac{1530}{T}\ \text{kJ/kmol.K} \qquad \textbf{(232,559.55 kJ/kg)}$$

4. An engine cylinder has a piston area of 0.12 m^2 and contains gas at a pressure of 1.5 MPa. The gas expands according to a process, which is represented by a straight line on a pressure-volume diagram. The final pressure is 0.15 MPa. Calculate the work done by the gas on the piston if the stroke is 0.3 m. **(23.4 kJ)**

5. 10 kg mol of a gas occupies a volume of 603.1 m^3 at a temperature of 140°C while its density is 0.464 kg/m^3. Find its molecular weight and gas constant and its pressure. **(28,296.9 J/kg.K)**

2

LAWS OF THERMODYNAMICS

2.1   INTRODUCTION

The laws of thermodynamics define the conversion of energy from one form to another. The first law of thermodynamics deals with the conservation of energy. It discusses the irreversible and reversible, steady and unsteady processes. It expresses the equations of energy for various types of engineering devices. The applications of second law are also discussed.

2.2   ZEROTH LAW

If there are three systems in thermal equilibrium and two of them have the same temperature, these two systems will have same temperature as that of the third one.

The idea of thermal equilibrium or temperature has been given by the zeroth law of thermodynamics.

The basis for temperature measurement is also given by the same law.

If the temperatures of two bodies 1 and 2 are compared with the help of a third body 3, the temperatures of bodies 1 and 2 are same even in the absence of thermal contact between them. Thermometer refers to body 3.

2.3   FIRST LAW OF THERMODYNAMICS

2.3.1 Definition

Energy can only be transformed, or changed from one form to another. It cannot be generated or crushed. Therefore, energy can neither be created nor destroyed. Total energy is conserved.

$$\text{Total energy}\,(E) = \text{Work}\,(W) + \text{Heat}\,(Q)$$

First law of thermodynamics gives the foundation for relations among the different types of energies and interactions of energies.

2.3.2 Energy Conservation

The total energy of the system is denoted by E. The unit of energy is J. It contains the following types of energy forms:

- thermal,
- kinetic,
- electrical,
- chemical,
- potential, and
- magnetic energies.

2.3.3 Energy Balance

The difference between the total energy entering the system and the total energy leaving the system is equal to the change in total energy of the system.

Total energy entering the system – Total energy leaving the system
= Change in the total energy of the system

Or,

$$E_{in} - E_{out} = \Delta E_{system}$$

This relationship is called the equation of energy balance and is applicable to all types of systems undergoing any process.

Change of energy of a system due to a process can also be calculated by differences between energy at the initial and at the final states.

Change of energy = E final state = E initial state.

Or,

$$\Delta E_{system} = E_{final} - E_{initial} = E_2 - E_1$$

If the state of system does not change by the process, change of energy of the system will be zero.

Different kinds of energies are prevailing some of which are internal (chemical, latent, nuclear, and sensible), magnetic, electric, potential, and kinetic. If surface tension, magnetic, and electric effects are absent, the total energy of a system will be the sum of its potential, kinetic, and internal energies.

$$\underbrace{E}_{\text{Total energy}} = \underbrace{U}_{\text{Internal energy}} + \underbrace{KE}_{\text{Kinetic energy}} + \underbrace{PE}_{\text{Potential energy}}$$

Or

$$\Delta E = \Delta U + \Delta KE + \Delta PE$$

where $\Delta U = m(u_2 - u_1)$

$$\Delta KE = \frac{1}{2}m\left(V_2^2 - V_1^2\right)$$

$$\Delta PE = mg(z_2 - z_1)$$

Figure 2.1 represents the first law of thermodynamics.

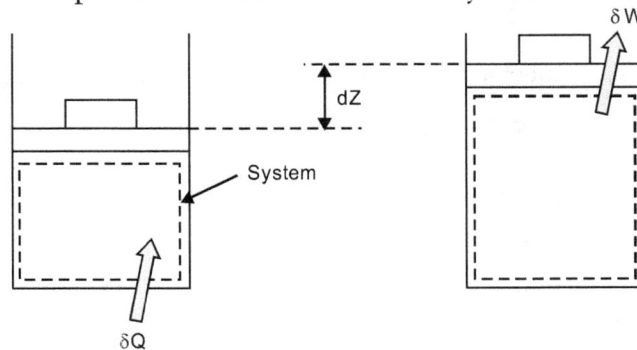

FIGURE 2.1 First law of thermodynamics

$$E_2 - E_1 = U_2 - U_1 + \frac{1}{2}m\left(V_2^2 - V_1^2\right) + mg(z_2 - z_1)$$

$$\therefore \quad E_2 - E_1 = U_2 - U_1 + \frac{1}{2}m\left(V_2^2 - V_1^2\right) + mg(z_2 - z_1) = {}_1Q_2 - {}_1W_2$$

For unit mass,

$$e_2 - e_1 = u_2 - u_1 + \frac{1}{2}\left(V_2^2 - V_1^2\right) + g(z_2 - z_1) = {}_1q_2 - {}_1w_2$$

- The internal energy per unit mass, or specific internal energy $= \dfrac{U}{m}$ (kJ/kg).

- The transfer of heat per unit mass, ${}_1q_2 \equiv {}_1Q_2 / m$ (kJ/kg).

- The work per unit mass, ${}_1w_2 \equiv {}_1W_2 / m$ (kJ/kg).

Most system has zero elevation or velocity. Therefore, changes in potential and kinetic energies are 0 (i.e., $\Delta PE = \Delta KE = 0$)

$$\therefore \quad \Delta E = \Delta U$$

EXAMPLE 2.1

A system of mass m has $d(PE) = 0, \delta Q = 0$, and $dU = 0$. This is a system with fixed internal energy, held at stationary height, and there are no heat interchanges with its surroundings. The first law gives equilibrium between variations in work and kinetic energy. The system is being speeded by a straight force F through a distance x. There is no force of friction. The system can be shown by Figure 2.2.

Solution: The first law of thermodynamics is given by:

$$-\delta W = d(KE)$$

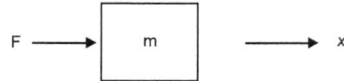

FIGURE 2.2 System speed by a force

Newtonian mechanics is expressed as follows:

$$-F \, dx = \delta W$$

Work is done by the system on the surrounding, hence the sign of work is negative.

$$\therefore \quad F \, dx = d(KE)$$

Newton's second law of thermodynamics shows that $m\left(\dfrac{d_2 x}{dt^2}\right) = F$

As $v = \dfrac{dx}{dt}$, so $m\left(\dfrac{dv}{dt}\right) = F$,

$$m\left(\frac{dv}{dt}\right)dx = d(KE)$$

$$v = \text{velocity}$$

Divide both sides by dt,

$$m\frac{dv}{dt}\frac{dx}{dt} = \frac{d(KE)}{dt}$$

$$m\frac{dv}{dt}v = \frac{d(KE)}{dt}$$

$$= vm\frac{dv}{dt}$$

$$= m\frac{d}{dt}\left(\frac{v^2}{2}\right)$$

$$= \frac{d}{dt}\left(m\frac{v^2}{2}\right)$$

$$KE_2 - KE_1 = \frac{1}{2}m\left\{\left(v_2^2\right) - \left(v_2^1\right)\right\}$$

Assume that $v_1 = 0$ and the kinetic energy is zero at this position,

$$\text{K.E} = \frac{1}{2} m \left(v^2 \right)$$

EXAMPLE 2.2

Cold water is heated from 15°C to 75°C by steam at 300°C and 500 kPa for home hot water supply. If there is a condition that the steam must not condense, what is the quantity of steam required per kg of liquid water?

Solution:

There is transfer of heat from steam into liquid water. No work is done.

Energy equation for water is given by:

$$Q + h_i m_{liq} = h_e m_{liq} \Rightarrow m_{liq} \left(h_i - h_e \right) = Q$$

Properties of the liquid water can be taken from steam tables.

$$\Delta h_{liq} = -\left(h_i - h_e \right) = -\left(62.9800 - 313.9100 \right) = 250.9300 \text{ kJ} / \text{kg}$$

$$\left(\cong C_p \, \Delta T = 4.1800 \, (75.00 - 15.00) = 250.800 \text{ kJ} / \text{kg} \right)$$

Energy of steam line has identical transfer of heat but it comes out of the steam

Energy equation for steam: $m_{steam} \, h_i = m_{steam} \, h_e + Q \Rightarrow -m_{steam} \left(h_e - h_i \right) = Q$

For properties of the steam, see steam table at a pressure of 500 kPa

$$\Delta h_{steam} = -\left(h_e - h_i \right) = -\left(2748.6700 - 3064.200 \right) = 315.5300 \text{ kJ} / \text{kg}$$

$$\therefore \frac{m_{liq}}{\Delta h_{steam}} = \frac{\Delta h_{steam}}{\Delta h_{liq}}$$

$$= \frac{250.93315.53}{315.5300} = \textbf{0.795 kg of steam per kg of water Ans}$$

2.3.4 Mechanisms of Energy Transfer, E_{in} and E_{out}

Transfer of energy from or to a system can take place in three forms: mass flow, heat, and work. Work and heat transfer represents only two forms of energy transfer in a closed system of constant mass.

1. **Transfer of Heat:** Heat transfer to a system (heat earn) raises the molecular energy and hence the internal energy of the system, and heat transfer

from a system (heat loss) decreases the energy of the molecules of the system and hence the internal energy of the system.

2. **Transfer of Work:** When work is supplied to a system, it raises the system's energy (i.e., work transfer from a system). The internal energy of the system changes accordingly.

3. **Flow of Mass:** Mass flow represents an extra mechanism of transfer of energy. Because mass carries energy with it, thus when mass goes into a system, the energy confined within the system drops because the exit mass carries out some of the energy with it.

2.4 FIRST LAW OF THERMODYNAMICS FOR A CLOSED SYSTEM

For adiabatic closed systems, transfer of heat is zero. The transfer of work is zero for systems with no interaction of work. The energy transfer with mass is zero for systems that have no flow of mass through their boundary (i.e., close systems). A simple closed system is shown in Figure 2.3, where all the three possibilities are specified.

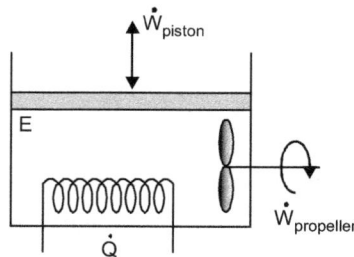

FIGURE 2.3 Closed system with energy E replacing heat Q and work W with its surrounding

Let U_1 be the internal energy of a closed system in equilibrium at state 1. If a quantity of heat $_1Q_2$ is transported through its boundary and an amount of work $_1W_2$ is applied by the system and the system again is in equilibrium state 2, then,
$$U_2 - U_1 = {_1Q_2} - {_1W_2}$$

If u is the specific internal energy of the system and m is the mass of the system then,

$$m\left({_1q_2} - {_1w_2}\right) = m\left(u_2 - u_1\right)$$

Or $$u_2 - u_1 = {_1q_2} - {_1w_2}$$

where $_1w_2$ and $_1q_2$ are work and heat transfer per unit mass of the system.

2.5 FIRST LAW OF THERMODYNAMICS FOR AN OPEN SYSTEM

(i) Flow Work

In an open system, fluid enters and leaves the system. The flow work of the fluid is needed to push to the system against the pressure of the system. To eject the fluid from the system, flow work is the product of specific volume, v and pressure, p.

$$\text{Flow work} = pv$$

(ii) Enthalpy

In the study of open systems, it is convenient to add the flow work "pv" with internal energy "u" to raise the system energy. The specific enthalpy h of the system can be defined as the total of flow work and internal energy. Therefore, specific enthalpy h is given as:

$$h = p.v + u$$

The mass crosses the boundary of an open system in addition to interaction of work and heat with the surroundings. The flow of fluid carries kinetic and potential energies.

In Figure. 2.4, V_2 and V_1 are the outlet and inlet velocities, m_2 and m_1 are the rates of mass flow at outlet and inlet, h_2 and h_1 are the enthalpies at outlet and inlet and z_2 and z_1 are the outlet and inlet elevations with reference to a datum; w and q are the rates of work and heat transfer to the system. Total energy of the system is E.

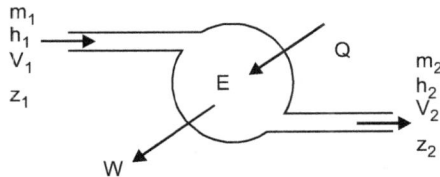

FIGURE 2.4 First law of thermodynamics of the open system

Energy can be transported in different mode such as mass, heat, and work. The net transfer of energy is the difference between the quantities transported at inlet and outlet. The energy equation becomes as follows:

$$\left(W_{out} - W_{in} \right) - \left(Q_{out} - Q_{in} \right) = E_{in} - E_{out}$$

where the subscripts "out" and "in" indicate amounts at exit and inlet of the system, respectively.

The first law of thermodynamics for a control volume can be defined as follows:

$$\underbrace{\frac{dE_{cv}}{}}_{\text{rate of } cv \text{ energy change}} = \underbrace{Q_{cv}}_{\text{CV heat transfer rate}} - \underbrace{W_{cv}}_{\text{CV shaft work rate}}$$

$$+ \underbrace{\sum m_i \left(h_i + \frac{1}{2} V_1^2 + gz_i \right)}_{\text{total enthalpy rate in}} - \underbrace{\sum m_e \left(h_e + \frac{1}{2} V_e^2 + gz_e \right)}_{\text{total enthalpy rate out}} \qquad 2.1$$

Total enthalpy is explained as:

$$h_{tot} = h + \frac{1}{2} V^2 + gz = u + \frac{1}{2} V^2 + gz + pv$$

The total enthalpy is the sum of kinetic, potential, internal, and flow work.

From Eq. (2.1),

$$\frac{dE_{cv}}{dt} = Q_{cv} - W_{cv} + \sum m_i h_{tot}, i - \sum m_e h_{tot}, e$$

or,

$$\frac{dE_{cv}}{dt} = Q_{cv} - W_{cv} + m_i \left(h_i + \frac{1}{2} V_i^2 + gz_i \right) - m_e \left(h_e + \frac{1}{2} V_e^2 + gz_e \right)$$

where i is at inlet and e at exit of the system, cv is control volume.

For steady state, $\dfrac{dE_{cv}}{dt} = 0$

$$\therefore \quad 0 = Q_{cv} - W_{cv} + m_i \left(h_i - h_e + \frac{1}{2} (V_i^2 - V_e^2) + g(z_i - z_e) \right)$$

$$\text{or,} \quad 0 = q - w + h_i - h_e + \frac{1}{2}(V_i^2 - V_e^2) + g(z_i - z_e)$$

$$\text{or,} \quad q + h_i + \frac{1}{2} V_i^2 + gz_i = w + h_e + \frac{1}{2} V_e^2 + gz_e$$

Rearranging the above equation, first law of thermodynamics becomes as follows:

$$\underbrace{h_e + \frac{1}{2} V_e^2 + gz_e}_{\text{outlet}} = \underbrace{h_i + \frac{1}{2} V_i^2 + gz_i}_{\text{inlet}} + \underbrace{q - w}_{\text{CV heat and work}}$$

For a steady-state open system

For steady-state steady-flow process, mass is constant, and the time rate of change of all quantities becomes equal to 0. The total energy of the system and the mass do not change with time, term dE/dt is 0 and for constant mass, $m=m_2=m_1$, the energy equation will be as follows:

$$\left(h_2 + \frac{1}{2}V_2^2 + gz_2\right) - \left(h_1 + \frac{1}{2}V_1^2 + gz_1\right) = q - w \qquad 2.2$$

where w and q are work and heat rate.

Eq. (2.2) is the equation for first law of thermodynamics.

2.6 SECOND LAW OF THERMODYNAMICS

First law of thermodynamics does not indicate the degree of transformation of cyclic heat into cyclic work. It indicates the possibilities of operations that can happen in nature, which preserve total energy. Clearly it is not the entire story concerning energy. As per first law, there will be no deficiency of energy as energy can be recycled.

First law of thermodynamics creates equality between the amount of the mechanical work and heat applied but does not identify the state conditions, under which transformation of heat into work is probable and not the direction in which transfer of heat can happen. The deficiency has been overcome by the second law of thermodynamics. The second law limits the machines from transforming whole of heat into work.

Entropy S can be defined as a thermodynamic property, which determines the amount of energy destruction.

2.6.1 Heat Engine

In a cyclic process, the initial state of a system is equal to the final state. The total change of the internal energy U will be zero.

$$\Delta U = 0$$

$$\therefore \ \ \Delta W = -\Delta Q$$

A process is reversible if the heat supplied is equal to the work produced by the system through a cycle. The work is negative if it is supplied to the system.

$$-\Delta W = \int P \, dV = \text{area enclosed}$$

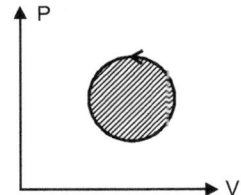

FIGURE 2.5 Work

A cyclic process can be regarded as a heat engine. Consider a heat engine operating between $T_1 > T_2$.

If Q is the heat supplied to the engine from a heat reservoir at temperature T_1, Q_1, is converted into work W, Q_2 represents the part of heat that is delivered to a second reservoir at $T_2 < T_1$ (condenser).

According to the first law of thermodynamics:

$$|Q_1| - |Q_2| = |W|$$

A heat engine is shown in Figures. 2.6 and 2.7. There are different types of heat engines. They undergo the following processes.

1. Heat engines take heat from a high temperature heat source (nuclear reactor, oil furnace, boiler, etc.)

2. Part of this heat is converted into work

3. The remaining heat is thrown into a low-temperature reservoir (condenser, atmosphere, lake, river)

4. Engine works in a cycle.

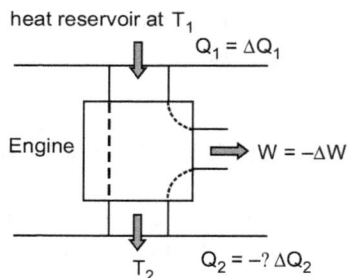

FIGURE 2.6 Heat engine

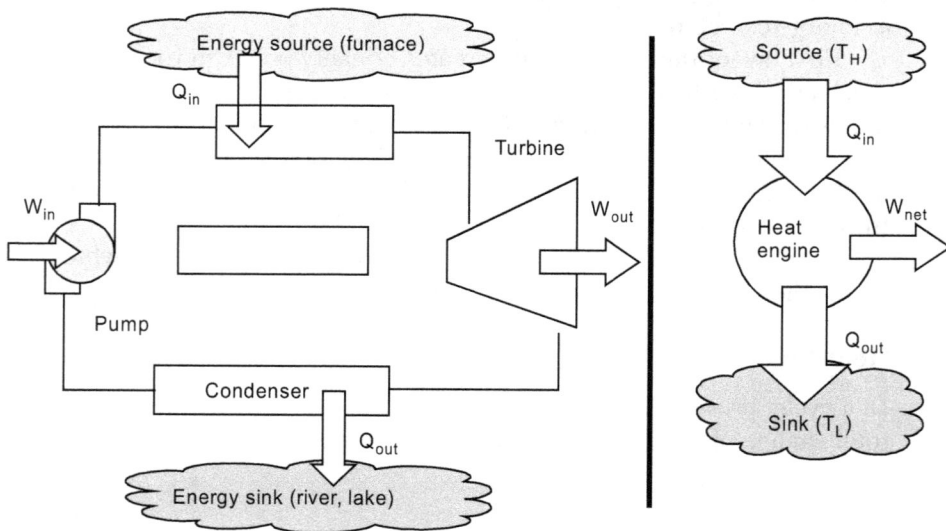

FIGURE 2.7 Heat engine of steam power plant

The ratio of work produced to heat supplied is called efficiency.

Efficiency = advantage/outlay (cost)

$$\eta_{th} = \frac{W_{net.out}}{Q_{in}} \text{ and } W_{(net,\,out)} = Q_{in} - Q_{out}$$

$$\eta_{th} = 1 - \frac{Q_{out}}{Q_{in}}$$

The engines have low thermal efficiencies. For example, engines of steam power plants have a thermal efficiency nearly 40%, sparkle-ignition motor car engine has about 20%, and diesel engines have roughly 30%.

The heat Q_{out} in power cycle cannot be saved. Without cooling in the condenser, the power cycle cannot be completed. All heat engines (even an ideal cycle) must reject some energy to basin at lower temperature so as to complete the cycle.

2.6.2 Heat Pumps and Refrigerators

Heat passes from areas of high temperature to ones of low temperature. The reverse cannot happen passively. Because such transfers need special machines called refrigerators.

Refrigerators are cyclical engines, and the fluids used in the cycles are called refrigerants. Heat pumps are engines that work to transfer heat from a medium at low temperature to a region at higher temperature.

Heat pumps and refrigerators are shown in Figure 2.8. These are basically similar machines but with different purpose only. Refrigerator maintains a region at a low temperature, whereas a heat pump takes heat from a region at low temperature and supplies it to a warmer region.

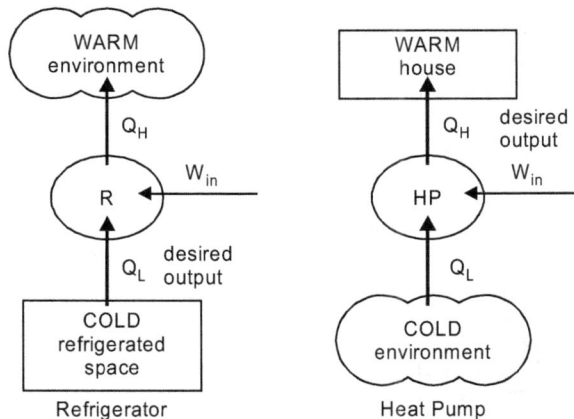

FIGURE 2.8 Heat pump and the refrigerator

2.6.3 Coefficient of Performance (COP)

The coefficient of performance (COP) can be defined as the performance of heat pumps and refrigerators that is expressed by the following formula:

$$COP_{HP} = \frac{\text{Benefit}}{\text{Cost}} = \frac{q_H}{w_c}$$

$$COP_R = \frac{\text{Benefit}}{\text{Cost}} = \frac{q_L}{w_c}$$

$$COP_{HP} = COP_R + 1$$

It can be realized that air conditioners are refrigerators, which cool the regions such as a building or a room.

The quantity of heat removed from the cooled region in BTU's units for one Watt-hour is called the Energy Efficacy Rating (EER).

$$3.41200 \ COP_R = EER$$

In general, air conditioners have an EER in the range from 8.00 to 12.00 (Coefficient of Performance 2.300 to 3.500).

2.6.4 Second Law of Thermodynamics: Clarification of Clausius Statement

It is not possible for a machine that works in a cycle to transfer heat from a body at lower temperature to a body at higher temperature without the aid of external work. In other words, a refrigerator can work only when its compressor is supplied with external power. Clausius and Kelvin-Planck statements of the second law of thermodynamics are contrary to each other. The two explanations of the second law are similar as presented in Figure 2.9. In other words any machine that contradicts the statement must conduct as that of Kelvin-Planck and Clausius and vice-versa.

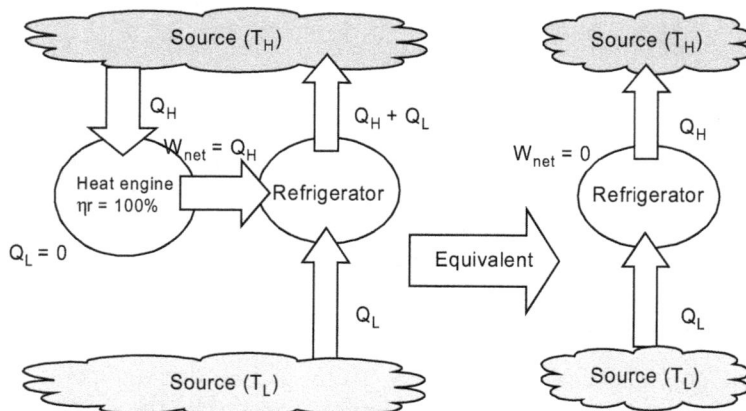

FIGURE 2.9 Clausius statement and Kelvin-Planck statement

Any machine that contradicts the second law of thermodynamics is called a **perpetual motion machine** of the second type (PMM2) and the machine that contradicts the first law of thermodynamics (by generating energy) is called a perpetual motion of the first type (PMM1).

EXAMPLE 2.3

A heat engine is supplied with 2512 kJ/min of heat at 650°C. Heat rejection takes place at 100°C. Specify the following results of heat rejection.

    (i) 867 kJ/min
    (ii) 1015 kJ/min
    (iii) 1494 kJ/min

Solution: The schematic diagram of the engine is shown in Figure 2.10

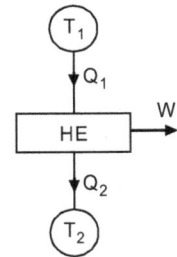

FIGURE 2.10 Heat engine

$$T_1 = 650°C + 273 = 923\,K$$
$$T_2 = 100°C + 273 = 373\,K$$
$$Q_1 = 2512\ kJ/min$$

(i) Q_2 = 867 kJ/min
Applying Clausius inequality

$$\frac{Q_1}{T_1} - \frac{Q_2}{T_2} = \frac{2512}{923} - \frac{867}{373} = 0.34770 > 0$$

The cycle is not possible.
(ii) Q_2 = 1015 kJ/min

$$\frac{2512}{923} - \frac{1015}{373} = 0$$

The cycle is reversible
(iii) Q_2 = 1494 kJ/min

$$\frac{2512}{923} - \frac{1494}{373} = -1.284 < 0$$

The cycle is irreversible and possible.

REVIEW QUESTIONS

    1.   What is zeroth law of thermodynamics? How does the concept of thermal equilibrium lead to invention of thermometer?

2. State the First Law of Thermodynamics. Derive energy equation in different forms.

3. Define:

    (i) Total energy

    (ii) Flow energy

    (iii) Enthalpy

4. Define the conditions for application of First Law of Thermodynamics to a closed system.

5. Derive energy equation for steady-state steady-flow open system.

6. Why is the Second Law of thermodynamics needed?

7. Define engine efficiency and coefficient of performance (COP) of a heat pump and a refrigerator. Prove that COP of a heat pump is more than that of a refrigerator.

8. Explain Perpetual Motion Machines of First Type and Second Type.

NUMERICAL EXERCISES

1. A reversible heat engine operates between reservoirs at 420 and 280 K. If the engine output is 2.5 kJ, determine the efficiency of the engine and its heat interactions with the two reservoirs. Subsequently, the engine is reversed and made to operate as heat pump between the same reservoirs. Calculate the coefficient of the heat pump and power input required when the heat transfer rate from the 280-K reservoir is 4 kW.

    (5 kJ, 2.5 kW)

2. Obtain the coefficient of performance of the composite refrigerator system in which two reversible refrigerator B only.

$$\left[R_{12} = \frac{R_1 R_2}{(1 + R_1)(1 + R_2) - R_1 R_2} \right]$$

3. A Carnot engine E_1 operates between temperatures T_1 and T_2 and engine E_2 operates between temperatures T_2 and T_3 receiving heat rejected from engine E_1. What is the relationship of η_2 with η_3 of engine E_3 working between T_1 and T_3

    ($\eta_3 = \eta_1 + \eta_2 - \eta_1 \eta_2$)

4. A cold storage of 100 tonnes of refrigeration capacity runs at 1/4th of its Carnot coefficient of performance. Inside temperature is –15°C and atmospheric temperature is 35°C. Determine the power required to run the plant. Take one metric ton of refrigeration as 3.52 kW.   **(281.6 kW)**

5. The steam supply to an engine comprises two streams, which mix before entering the engine. One stream is supplied at the rate of 0.01 kg/s with the enthalpy and a velocity of 20 m/s. The other stream is supplied at the rate of 0.1 kg/s with an enthalpy of 2665 kJ/kg and a velocity of 120 m/s. At the exit from the engine, the fluid leaves as two streams, one of water at the rate of 0.001 kg/s with an enthalpy 421 kJ/kg and the other of stream. The fluid velocities at the exit are negligible. The engine develops a shaft power of 25 kW. The heat transfer is negligible. Estimate the enthalpy of second exit steam.   **(2489 kJ/kg)**

SECOND LAW OF THERMODYNAMICS AND ENTROPY

3.1   INTRODUCTION

Entropy might give an indication about the second law of thermodynamics. The concept of entropy is as fundamental to physics as energy or temperature. However, it is a distinct concept related to the amount of disorder of a system. As the disorder of the system increases, the entropy increases. The system's molecular motion increases with the increase in thesystem's internal energy. Heat addition increases the system disorder by increasing the molecular motion randomness. In general, heat addition increases the system disorder. However, heat addition effect on the cold (low temperature) and hot (high temperature) systems is not the same. In other words, for the same amount of the added heat, disorder level in the cold systems is much higher than that of the hot systems. The entropy change for a reversible system, which can be defined as the system in which the variation in the entropy takes place due to the variation in the system's internal energy, can be calculated as shown in Equation 3.1.

$$\text{Change in entropy} = \frac{\text{Change in the heat of system}}{\text{Temperature}}$$

$$\text{Or,} \qquad \Delta S = \frac{\Delta Q}{T} \tag{3.1}$$

where ΔS, ΔQ, and T represents the entropy change (joule/Kelvin cr calories/Kelvin), heat change of the system (Joules or calories), and temperature (Kelvin), respectively.

Second law of thermodynamics may be simplified as follows:

1. The entropy of the system will naturally increase or there will be no change in the system's entropy if the system is already maintained at the maximum level of entropy.

2. If the system is left to itself, it naturally tends to be in equilibrium within the surroundings.

3. When the system and the surroundings are maintained in equilibrium, there is no changes in the entropy of a system.

For a real process, energy quantity is preserved; however, it changes from one form to another and its quality is certain to decrease. Disorganization is the main result of the process due to the changes in the energy levels and in the forms. Therefore, disorganized energy measurement and the entropy terms are not disconnected. In other words, the terms entropy, energy, and exergy are inter-dependent, and one affects the other in some way as shown in Figure 3.1.

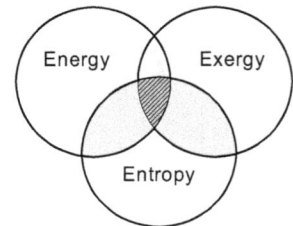

FIGURE 3.1 Interaction between the energy, the exergy, and the entropy.

Figure 3.1 describes the relationship between these three terms so that the unknown can be easily obtained if the other two are known. This can be expressed as follows:

$$\text{(Entropy balance)} = \text{(Energy balance)} - \text{(Exergy balance)} \qquad (3.2)$$

3.2 REVERSIBLE AND IRREVERSIBLE PROCESSES

According to the second law of thermodynamics, a heat engine cannot achieve 100% efficiency. Then, what is the maximum efficiency that can be achieved by a heat engine? It depends on the following two types of processes: reversible process (an ideal process) and irreversible process. The reversible process can be defined as the process in which the system and the surrounding (both of them) can return to their original state (return of process is possible). Whereas, the irreversible process is defined as the process in which the system and the surrounding (both of them) cannot return to their original state (return process is impossible). Restoring both the system and the surrounding together is the necessary condition for the reversible process. In other words, if any one of them failed to restore back to their original state while the other did, it is said to be an irreversible process. The reversible process is only an assumption for ideal processes, which is difficult to achieve. With certain assumptions, the systems can be approximated to get reasonable results. Quasi-static process is used for two reasons: (a) it is easier for the analysis as long as the system involves several equilibrium states, and (b) it works as the boundaries (limits) for comparison with the actual process. No reversible process is possible to happen in the nature. All the processes happening in the nature are irreversible processes.

The reversible process cannot be accomplished, but the processes are designed to get closer. The closer the process, the more output work delivered or the minimum heat input consumed by the system.

The concept of reversible process and the definition of efficiency of the second law of thermodynamics are not independent for the actual process. System efficiency allows to compare between different machines (designed to do the same task) based on the comparison between their efficiencies. The machines should be designed for the lowest level of irreversibility.

3.2.1 Irreversibility

Irreversibility can be defined as the factors that shift any process from reversible to irreversible process. Some of the good examples of these are heat transfer, frictional force, chemical reactions, electric resistance, and an inelastic deformation of solids. Any one of these factors in a process can lead to irreversibility.

3.2.2 Friction

Friction can be defined as the resistance to the motion for two or more bodies in motion. It is a very familiar form of irreversibility. The frictional force can be developed between any two surfaces in contact, such as the force between the surfaces of a piston moving inside a cylinder. This force consumes some amount of system's energy and therefore, specific amount of the work is required to substitute the reduction in the system's energy. In the piston–cylinder example, the frictional energy can be recognized by the heat developed between the two surfaces in contact. Temperature increasing at the same surfaces is another indication about friction. The developed heat does not depend on the direction of motion. Heat develops whatever the direction of motion is. This process can be considered as an irreversible process because the system and its surrounding are not able to return to their original state (irreversible process). The degree of irreversibility varies with the value of force of friction and is linearly dependent.

The force of friction results at the interface between two solid bodies. It can also develop at the interface between a fluid and a solid and even between the layers of the fluid through a fluid flow due to velocity difference. The drag force of rigid surface motion in air is a good example of force of friction that cannot be overcome without applying an external source of work and energy.

3.2.3 Factors Causing an Irreversible Process

There are several factors which make a process irreversible. They are as follows:
- the chemical action and reaction,
- process involves mixing,
- heat transfer,
- unrecovered deformation,
- compression and unrestrained expansion and compression.

3.2.4 Internal and External Reversibilities

The internal irreversibilities occur within the system boundaries. The system may be at several states of equilibrium. It will pass through all these states when returned to its original state.

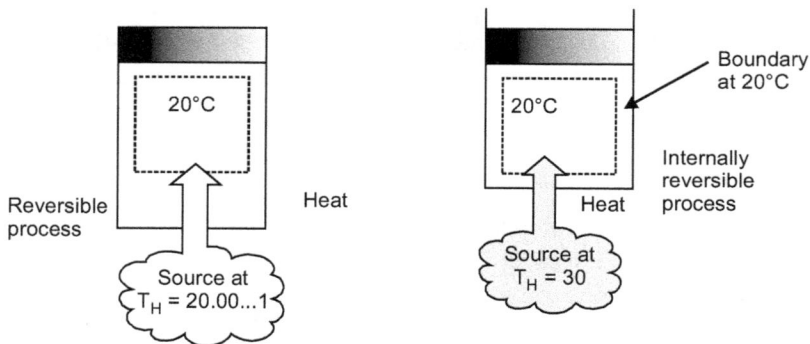

FIGURE 3.2 Reversible and internally reversible processes.

The external irreversibilities occur outside the system boundaries. For example, heat transfer occur between a system and a reservoir both at the same temperature. In a totally reversible process, both internal and external irreversibilities should be absent.

3.3 ENTROPY

Calculation of entropy requires information about the number of states of molecules energy. However, in classical thermodynamics, entropy can be estimated in terms of properties at the macroscopic level.

3.3.1 Mathematical Definition

For an internally reversible process, the condition $\oint \left(\dfrac{\delta Q}{T} \right)_{int\ rev.} = 0$ must be valid.

Therefore, for a reversible process,

$$\left(\frac{\delta Q}{T} \right)_{rev} = dS.$$

For convenience, the subscript "*int*" is ignored.

3.3.2 Characteristics of Entropy

An alternative definition of entropy is an energy distribution measurement at the molecular levels of the matter involved in a system. Higher value of the entropy indicates larger number of levels of energy distribution. *T-s* diagram can be used to depict the processes as shown in Figure 3.3. The amount of heat transferred (reversible heat) can be represented by the area under the curve's connecting points 1 and 2 on the *T-s* diagram. For the adiabatic and reversible process, the term $\delta Q_{rev} = 0$ indicates that there is no change in the entropy during the process.

Figure 3.4 shows the Carnot cycle on *T-s* diagram. The rectangular representation of the cycle means that changes in the entropy of both heat absorption (ΔS_H) and heat rejection (ΔS_L) are the same. However, change of entropy for the total system ΔS can be calculated by adding the changes of the entropies of the subsystems together.

$$\therefore \quad \Delta S_{1+2} = \Delta S_1 + \Delta S_2$$

$$\text{Or} \quad S_{1+2} = S_1 + S_2$$

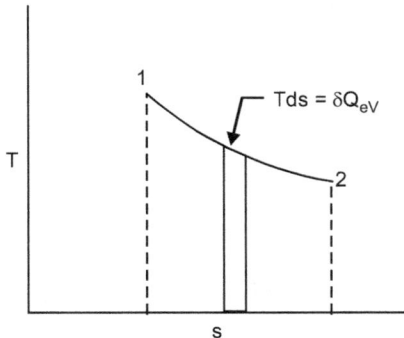

FIGURE 3.3 A process represented on the temperature–entropy (*T*-s) diagram.

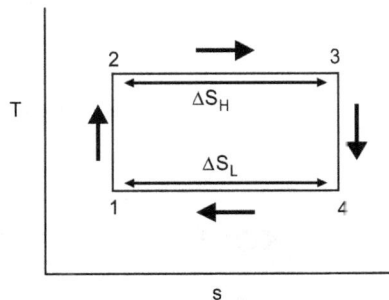

FIGURE 3.4 A Carnot cycle represented on a *T*-s diagram.

3.3.3 Entropy as a Thermodynamic Property

In a closed system, the first law of thermodynamics of a reversible process can be written as follows:

$$\delta U = \delta Q + \delta W$$
$$\delta U = T \, dS - P \, dV$$
$$\delta Q = T \, dS$$

This equation shows that the entropy S and the volume V are the two state variables for which the differential of the internal energy $U = U(S, T)$ becomes exact. One basic pair of state variables will transform to another, namely from (S, V) to (T, V) when one of the variables (entropy S) takes the role of a thermodynamic property. The differential of the entropy can be derived by inserting dS and dU/T of the forms.

$$dS = \left(\frac{\delta S}{\delta T} \right)_V dT + \left(\frac{\delta S}{\delta T} \right)_T dV$$

$$dU = \left(\frac{\delta U}{\delta T} \right)_V dT + \left(\frac{\delta U}{\delta V} \right)_T dV$$

By substituting in Equation (3.3), it leads to

$$\left(\frac{\delta S}{\delta T} \right)_V dT + \left(\frac{\delta S}{\delta V} \right)_T dV = \frac{1}{T} \left(\frac{\delta U}{\delta T} \right)_V dT + \frac{1}{T} \left(\frac{\delta U}{\delta V} \right)_V dV + P) \, dV$$

where the terms $(\delta S / \delta T)_V = (\partial U / \partial T)_V$ is defined by C_V. Partial derivatives of the entropy, as a state function, with respect to T and V can be written as follow:

$$\left(\frac{\delta S}{\delta T} \right)_V = \frac{CV}{T} \tag{3.4}$$

$$\left(\frac{\delta S}{\delta V} \right)_T = \frac{1}{T} \left[\left(\frac{\delta U}{\delta V} \right)_V + P \right]$$

3.4 ENTROPY BALANCE FOR CLOSED SYSTEMS

3.4.1 Closed Systems

Heat transfer increases the entropy of the system (system disorder) according to the statement presented by Equation (3.5) so that,

$$\left\{ \begin{array}{l} \textit{Time rate of entropy accumulation} \\ \textit{within a system at an instant of time t} \end{array} \right\} =$$

$$\left\{ \begin{array}{l} \textit{The net rate of entropy transport} \\ \textit{into the system by heat transfer at} \\ \textit{an isothermal boundary of T} \end{array} \right\} + \left\{ \begin{array}{l} \textit{The time rate of entropy} \\ \textit{generation due to irreversibilities} \\ \textit{of the process} \end{array} \right\} \qquad (3.5)$$

The degree of irreversibility is the reason for the change of the entropy of an irreversible process. The entropy generation for an adiabatic system due to irreversible process can be given by:

$$S_{gen} = \frac{1}{T_0} \quad or \quad (S_2 - S_1)_{gen} = \frac{I_2}{T_0} \qquad (3.6)$$

Equation (3.6) states that entropy generation depends on the degree of the process irreversibility and the factor $1/T_0$. However, entropy generation will be zero ($S_{gen} = 0$) when the process is reversible and adiabatic. In this case, the process is called "isentropic process." Equation (3.6) shows that for an isolated system, the system entropy does not change if the process is a reversible process. For $(\Delta S)_{iso} \geq 0$, is always valid for an isolated system.

EXAMPLE 3.1

Find the relationship between the pressure and the volume of your process for an ideal gas in piston–cylinder device knowing that the process is isentropic.

Solution: Since the process is isentropic, the change of the entropy must be zero, that is $S_2 - S_1 = 0$. Therefore, $(P_2 / P_1)^{cv} = (v_1 / v_2)^{cp} \, (P_2 / P_1) = (v_1 / v_2)^{cp/cv}$. cp / cv is equal to k so that $k = cp / cv$. Hence, the process can be described by the expression $pv^k =$ constant.

EXAMPLE 3.2

For a substance flow inside your pipe (incompressible flow), derive a mathematical expression for the isentropic shaft work.

Solution: Flow inside your pipe may be affected by several parameters, such as a pump work to overcome the losses due to friction, pipe diameter and cross section, changes in pipe elevation, etc. Taking most of these parameters into consideration, the flow equation (energy equation) can be written as follows:

$$\left(h + \frac{1}{2}v^2 + gz \right)_i = q - w$$

For incompressible flow,

$$\Delta h = \Delta u + v\Delta p$$
$$-w_{shaft} = v\Delta p + \Delta ke + \Delta pe$$

For frictionless and incompressible flow, the fluid flows due to elevation, velocity, and pressure changes. Steady flow energy in the absence of work done by the pump can be given by,

$$v\Delta p + \Delta ke + \Delta pe = 0$$

Or

$$v(p_2 - p_1) + \frac{1}{2}(v_2^2 - v_1^2) + g(z_2 - z_1) = 0$$

The last equation, called the Bernoulli's equation, describes the isentropic flow inside a pipe for incompressible fluids.

3.4.2 Thermodynamic Cycles

The cyclic integral of entropy with respect to time may be expressed as follows:

$$\oint S dT = \oint \frac{\delta Q}{T} + \oint \frac{\delta I}{T_0} \tag{3.7}$$

As entropy is a system property, the integral (cyclic) of entropy of the occupied flared must be zero, that is $\oint S\,dt = 0$. For all cyclic processes, the irreversibility cannot be negative; $\oint dI \geq 0$, then, from Equation (3.7), Equation (3.8) must be valid for all possible cycles, that is

$$\oint \frac{dQ}{T} \leq 0 \tag{3.8}$$

The inequality shown in Equation (3.8) is called Clausius inequality. It was introduced by R.J.E. Clausius, German physicist in 1870. This inequality is used to test the validity of the second law of thermodynamics for a cycle. This inequality states that $\oint \frac{dQ}{T}$ is negative for the real cycles, while equals to zero for a reversible process.

EXAMPLE 3.3

A heat engine withdraws 325 kJ of heat energy drawn from a high temperature reservoir at 1000K and passes 125 kJ of the heat energy to a low temperature reservoir at 400K.

The required work output (net) of the engine is 200 kJ. Is that possible?

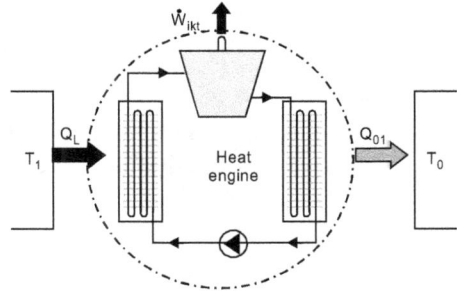

FIGURE 3.5 Validity of a heat engine

Solution: The first law of thermodynamics states that:

$$Q_{net} = W_{net}$$

Or $$W_{net} = 325 - 125 = 200\,\text{kJ}$$

It is possible for the engine to provide the required energy. However, this engine might not be valid in the light of the second law of thermodynamics. Application of Equation (3.7) yields,

$$\oint \frac{dQ}{T} = \frac{Q_1}{T_1} = \frac{Q_{01}}{T_0}$$

$$= \frac{325}{1000} - \frac{125}{400}$$

$$= 0.0125\ \text{kJ/K} > 0$$

Therefore, this engine is not possible.

3.5   ENTROPY BALANCE FOR OPEN SYSTEMS

For open systems, the net entropy transfer because of mass flow can be given as follows:

$$\left\{ \begin{array}{l} \textit{The net rate of entropy accumulation} \\ \textit{by mass flow at an instant of time, t} \end{array} \right\} = \sum_i \dot{m}_1 s_i - \sum_e \dot{m}_e s_e \ \ldots \quad (3.9)$$

In general, the rate of a system's entropy change can be calculated as follows:

$$\left(\frac{dS}{dt}\right)_{cv} = \int_A \left(\frac{\dot{q}}{T}\right)dA + \sum_i \dot{m}_1 s_i - \sum_e \dot{m}_e s_e + \dot{S}_{gen} \ldots \quad (3.10)$$

It is important to notice that system energy and system energy of the steady state steady flow (SSSF) system are constant. For this process, substituting $\left(\dfrac{dS}{dt}\right)_{cv} = 0$ into Equation (3.10) and with some mathematical manipulation results in,

$$\sum_e \dot{m}_e s_e = \int_A \left(\frac{\dot{q}}{T}\right)dA + \sum_i \dot{m}_i s_i + \dot{S}_{gen} \ldots \quad (3.11)$$

For an adiabatic SSSF of single input and single output system, Equation (3.11) can be modified into Equation (3.12) as follows:

$$S_e = S_i + S_{gen} \quad (3.12)$$

The quantity S_{gen} is a positive value due to fluid and surface friction. For the adiabatic SSSF, the inequality $S_e > S_i$ is always valid and true. This means that the losses in system energy correspond to gains in entropy through the process. In other words, the process at which both energy and entropy are conserved and showed no increase in value is called the principle of entropy increase.

EXAMPLE 3.4

Figure 3.5 shows duct of a ventilation system working in a well-insulated steady state system. The added pressure is 1 atm. Air with $c_p = 1.005$ kJ/kg·K as assumed a perfect gas flows inside the duct. By neglecting potential and kinetic energy, find:

    A. air temperature at the exit of the duct,
    B. pipe diameter at the exit of the duct,
    C. the entropy production rate in the duct measured as kW/K.

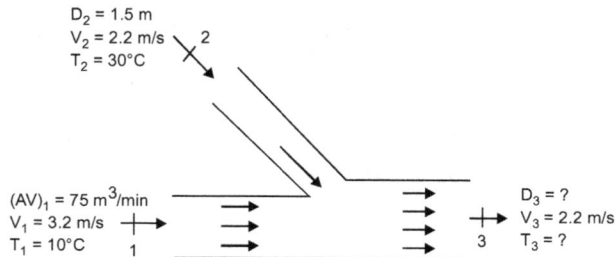

$D_2 = 1.5$ m
$V_2 = 2.2$ m/s
$T_2 = 30°C$
2

$(AV)_1 = 75$ m³/min
$V_1 = 3.2$ m/s
$T_1 = 10°C$
1

$D_3 = ?$
$V_3 = 2.2$ m/s
$T_3 = ?$
3

FIGURE 3.6 A ventilation system.

Solution:

A. The energy equation for the airflow inside the duct shown in Figure 3.6 can be written as,

$$\dot{m}_1 h_1 + \dot{m}_2 h_2 = \dot{m}_3 h_3.$$

The mass conservation yields, $\dot{m}_1 + \dot{m}_2 = \dot{m}_3$.

$$\dot{m}_1 = \rho_1 (AV)_1$$

$$\rho_1 = \frac{100}{0.287283} = 1.283 \text{kg/m}^3$$

$$\therefore \quad \dot{m}_1 = 1.603 \text{ kg/s}$$

Similarly, $\qquad \rho_2 = 1.149 \text{ kg/m}^3$

And $\qquad \dot{m}_2 = \rho_2 \dfrac{\pi D^2}{4} V_2$

After substituting the numerical values, $\dot{m}_2 = 4.464$ kg/s and $\dot{m}_3 = 6.067$ kg/s. For an ideal gas, $h = c_p T$. Using this expression in the equation of the energy above, air temperature at the exit can be calculated as,

$$T_3 = \frac{1.063 \times 282 + 4.464 \times 303}{6.067} = 297.7 \text{k}$$

B. Since $\dot{m}_3 = \rho_3 A_3 V_3$ and $\rho_3 = \dfrac{100}{0.287 \times 297.7} = 1.17 \text{ kg/m}^3$, the diameter of the circular cross section becomes $D_3 = 1.73$ m.

C. For the insulated steady state system,

$$S_{gen} = \dot{m}_1 (S_3 - S_1) + \dot{m}_2 (S_3 - S_2)$$

Using the ideal gas entropy relation, the entropy generated is equal to 2.412103 kW/K.

3.6 ENTROPY OF THE IDEAL GAS

The specific heat Cv and the free energy U of the ideal gas are given as follows:

$$Cv = \frac{3}{2} nR \ \text{ and } \ U = \frac{3}{3} nRT \qquad \left(\frac{\delta U}{\delta V} \right)_T = 0$$

Using the partial derivatives and the equation of the state $PV = nRT$ of the ideal gas leads to the expression

$$dS = \frac{3}{2}nR\frac{dT}{T} + nR\frac{dV}{V}$$

The entropy difference is given by Equation (3.13) as follows:

$$S(T,V) - S(T_0,V_0) = \frac{3nR}{2}\log\left(\frac{V}{V_0}\right) \tag{3.13}$$

where n is the number of moles (constant in this equation).

3.7 MAXWELL EQUATIONS

The commutativity of the differentiation operations, Schwarz's theorem, can be used to derive relations between thermodynamic quantities.

For the case of the differential, this implies that $\dfrac{\partial^2 U}{(\partial S\,\partial V)} = \dfrac{\partial^2 U}{(\partial V\,\partial S)}$

Hence, $\qquad dU = T\,ds - P\,dV, \quad \dfrac{\delta T}{\delta V} = -\dfrac{\delta P}{\delta S}$

$$\left(\frac{\delta V}{\delta T}\right)_p = -\left(\frac{\delta S}{\delta P}\right)_T \tag{3.14}$$

where an inversion for the last step was used. The relation $\left(\dfrac{\delta V}{\delta T}\right)_p = \left(\dfrac{\delta S}{\delta P}\right)_T$ is called Maxwell's equation.

3.7.1 Energy Equation

The entropy is not an experimentally controllable variable, such as T, V, and P, which allowed to the thermal equation measurement of the state $P = P(T, V)$.

However, Equation (3.1) is used to deduce the energy equation, which allows to determine the equation of state $U = U(T, V)$.

Energy equation: Derivatives of the entropy was used as:

$$\frac{\partial}{\partial V}\left(\frac{1}{T}\frac{\partial U}{\partial T}\right) = \frac{\partial}{\partial V}\left(\frac{\partial S}{\partial T}\right) = \frac{\partial}{\partial T}\left(\frac{\partial S}{\partial V}\right) = \frac{\partial}{\partial T}\left(\frac{1}{T}\left[\frac{\partial U}{\partial T} + P\right]\right)$$

$$= -\frac{1}{T^2}\left[\frac{\partial U}{\partial V} + P\right] + \frac{1}{T}\left[\frac{\partial^2 U}{\partial T\,\partial V} + \frac{\partial P}{\partial T}\right]$$

Cancelling identical terms leads to:

$$\left(\frac{\partial T}{\partial V}\right)_T = T\left(\frac{\partial P}{\partial T}\right)_v - P \tag{3.15}$$

This is called energy equation.

3.8 EULER'S CYCLIC CHAIN RULE

The partial derivatives $(\partial P / \partial T)_v$ in the energy equation (Equation (3.15)) may be replaced by thermodynamic coefficients. The involved type of variable transformation can be applied to a large set of thermodynamic quantities.

Implicate variable dependencies: In general, equation of state is described by a set of variables, example P, V, and T, so that:

$$f(P,V,T)=0, \quad \frac{\partial f}{\partial P}dP + \frac{\partial f}{\partial T}dT + \frac{\partial f}{\partial V}dV = df(P,V,T)=0$$

From the relative partial derivatives of the state variables,

$$\left(\frac{\partial P}{\partial T}\right)_V = -\frac{\dfrac{\partial f}{\partial T}}{\dfrac{\partial f}{\partial P}}, \quad \left(\frac{\partial T}{\partial V}\right)_P = -\frac{\dfrac{\partial f}{\partial V}}{\dfrac{\partial f}{\partial T}}, \quad \left(\frac{\partial V}{\partial P}\right)_T = -\frac{\dfrac{\partial f}{\partial P}}{\dfrac{\partial f}{\partial V}}$$

This relation can be rewritten as:

$$\left(\frac{\partial P}{\partial T}\right)_V \left(\frac{\partial T}{\partial V}\right)_P \left(\frac{\partial V}{\partial P}\right)_T = -1 \tag{3.16}$$

The importance of Equation (3.16), Euler's chain rule, lies in the fact that one does not know the equation-of-state function $f(P, V, T)$ explicitly.

Expansion and compression coefficients: Using Equation (3.16), we get:

$$\left(\frac{\partial P}{\partial T}\right)_V = -\frac{1}{\left(\dfrac{\partial T}{\partial V}\right)_P \left(\dfrac{\partial V}{\partial P}\right)_T} = -\frac{\left(\dfrac{\partial V}{\partial T}\right)_P}{\left(\dfrac{\partial V}{\partial P}\right)_T} \quad \left(\frac{\partial P}{\partial T}\right)_V = \frac{a}{kT} \tag{3.17}$$

The thermodynamic coefficients used are as follows:

$$\alpha = \frac{1}{V}\left(\frac{\partial V}{\partial T}\right)_P \quad \text{Coefficient of thermal expansion}$$

$$KT = -\left(\frac{\partial V}{\partial P}\right)_T \quad \text{Isothermal compressibility}$$

$$KS = -\frac{1}{V}\left(\frac{\partial V}{\partial P}\right)_S \quad \text{Adiabatic compressibility}$$

Using Equation (3.17), the energy equation (Equation (3.15)) leads to Meyer's relation between C_p and C_V.

$$\left(\frac{\partial U}{\partial V}\right)_T + P = T\left(\frac{\partial P}{\partial T}\right)_V = T\frac{\alpha}{KT} \tag{3.18}$$

The energy equation (Equation (3.18)) can be used to rewrite Mayer's relation, which can be derived by considering the chain rule for $(\partial U/\partial T)P$, as:

$$C_p = C_V + \left[P + \left(\frac{\partial U}{\partial V}\right)T\right]\left(\frac{\partial V}{\partial T}\right)_P$$

This leads to:

$$C_P - C_V = \frac{\alpha^2}{KT}TV > 0 \tag{3.19}$$

3.9 ENTROPY DIFFERENTIALS

The rewritten energy equation (Equation (3.18)) can be used to write the differential (Equation (3.20)) of the entropy as follows:

$$T\,dS = C_V\,dT + \left[\left(\frac{\partial U}{\partial V}\right)_T + P\right]dV$$

$$= C_V\,dT + T\frac{\alpha}{KT}dV \tag{3.20}$$

3.9.1 The Two Independent Variables, *T* and *V*

Equation (3.20) of the differential of the entropy implies that the absorbed heat δQ can be expressed likewise in terms of directly measurable coefficients,

$$\partial Q = T\,dS = C_V\,dT + \frac{\alpha}{KT}T\,dV \tag{3.21}$$

where T and V are the independent variables.

3.9.2 The Two Independent Variables, *T* and *P*

Using Maxwell's Equation (3.14),

$$\left(\frac{\partial V}{\partial T}\right)_P = -\left(\frac{\partial S}{\partial P}\right)_T$$

Entropy can be given in terms of *T* and *P* as,

$$dS = \left(\frac{\partial S}{\partial T}\right)_P dT + \left(\frac{\partial S}{\partial P}\right)_T dP$$

$$= \frac{C_P}{T} dT - \left(\frac{\partial V}{\partial T}\right)_P dP$$

$$\therefore \quad T\,dS = C_P\,dT - \alpha TV\,dV \qquad (3.22),$$

where *T* and *P* are the independent variables.

3.9.3 The Two Independent Variables, *V* and *P*

The term *dT* can be rewritten as

$$dT = \left(\frac{\partial T}{\partial V}\right)_P dV + \left(\frac{\partial T}{\partial P}\right)_V dP$$

$$= \frac{1}{\alpha V} dV + \frac{KT}{\alpha} dP$$

Using Equation (3.18), via $(\partial P/\partial T)_V = \alpha/KT$, and inserting *dT* into Equation (3.19) leads to:

$$T\,dS = \frac{C_P}{\alpha V} dV + \left(\frac{C_{PKT}}{\alpha} - \alpha TV\right) Dp \qquad (3.23)$$

For *T dS*, *V* and *P* are the two independent pairs of state variables.

3.10 TEMPERATURE–ENTROPY (*T-s*) DIAGRAM

The work due to system boundary movement can be expressed as follows:

$$W_{12} = \int_1^2 P\,dV,$$

where *P* is the applied pressure at the boundary. The temperature, pressure, and the other properties of a system involved in an ideal process must be uniform. If not, energy loss occurs because of the internal irreversibility. In this case, volume change of the system requires performing of an external work on the system equivalent to $\int_1^2 P \, dV$. The pressure in this equation is not the pressure applied at the boundary; but it is the pressure of the system.

Mathematically, the work of an ideal process can be calculated by the area under the p-V curve of the process shown in Figure 3.7. However, the work done for an irreversible process is smaller than the area under the curve for an ideal process. This is because of the non-uniform thermodynamic properties.

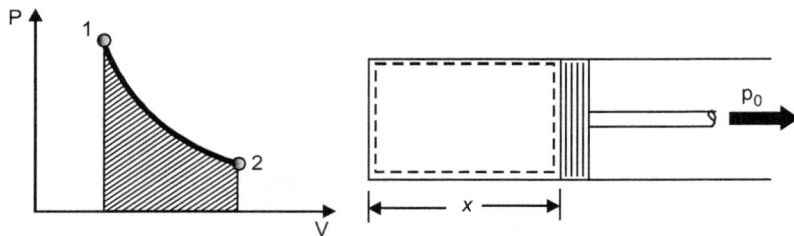

FIGURE 3.7 The mechanical work for an ideal process of a piston–cylinder system shown by the shaded area

Heat transfer for a reversible process in a closed system can be calculated by the shaded area under the T-s curve as shown in Figure 3.8 (a). Figure 3.8 (b) shows isentropic, isochoric, isobaric, and isothermal processes, on a T-s diagram.

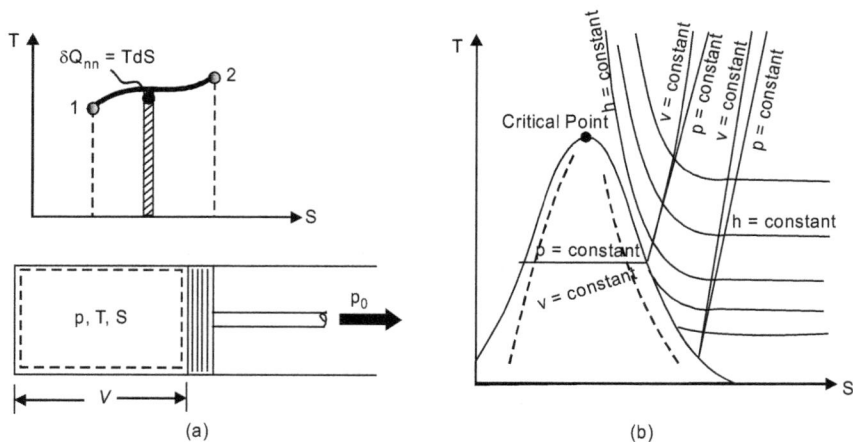

FIGURE 3.8 The ideal heat transfer to a system and T-s presentation of various ideal processes

3.11 ENTHALPY–ENTROPY (h-s) DIAGRAM

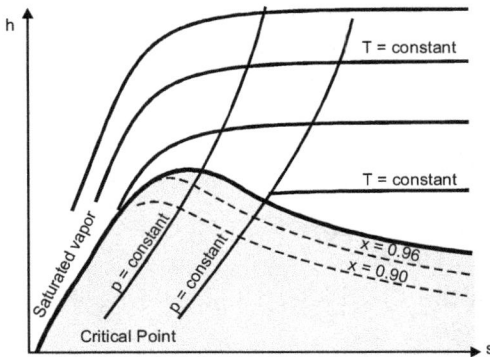

FIGURE 3.9 Mollier diagram

Enthalpy–entropy diagram for the SSSF process represents an important diagram.

The term "enthalpy" is an important factor in energy analysis of the systems with fluid flow because it is used to define the exergy of the system or the available energy in the system. Therefore, the processes involving exergy and energy interactions can be illustrated by h-s diagram. This diagram is called Mollier diagram developed by R dart Mollier, a German scientist. Figure 3.9 shows the general characteristics of a Mollier diagram.

On T-s diagram the constant pressure line is a horizontal line for a saturated state, while the constant pressure curves keep declining even for saturated states. In this graph, at a specific pressure, the temperature of the saturation can be found by the crossing point of constant temperature and constant pressure lines at the line of saturated vapor.

The enthalpy change, Δh, can be obtained by measuring the vertical distance between the required states on this diagram. This represents the consumed work of an adiabatic compressor or the work of the shaft of an adiabatic turbine. It can be the change in the kinetic energy of a fluid flows through a nozzle or a diffuser. On the other hand, the entropy generated Δs for a fluid flows adiabatically can be obtained by measuring the horizontal distance between two particular states on this diagram.

3.12 THIRD LAW OF THERMODYNAMICS (NERNST LAW)

Statistical study for more experiments done by Nernst showed that the entropy in the limit $T \to 0$ becomes a constant and independent of the other parameters, such a pressure and volume. This principle can be expressed as,

$$\left(\frac{\partial S}{\partial V}\right)_{T \to 0} = \left(\frac{\partial S}{\partial P}\right)_{T \to 0} = 0 \tag{3.24}$$

The entropy is defined as follows:

$$\lim_{T \to 0} S(T) = 0 \tag{3.25}$$

This equation is equivalent in statistical mechanics, where most of the states of matter are referred to a unique ground state. Macroscopically, at ground state leads to limit $T = 0$. Entropies are observed only for exotic phases of matter. Heat capacities vanish as $T \to 0$. The heat capacities disappear at $T = 0$ as a consequence of Equation (3.25) so that

$$\lim_{T \to 0} C_V = \lim_{T \to 0} T \left(\frac{\partial S}{\partial T} \right)_V = 0$$

$$\lim_{T \to 0} C_P = \lim_{T \to 0} T \left(\frac{\partial S}{\partial T} \right)_P = 0$$

(i) The ideal gas does not fulfil the third law: The heat capacities of the ideal gas are constants and can be given by,

$$C_V = \frac{3}{2} nR$$

$$C_P = \frac{5}{2} nR$$

This contradicts the third law which states that gas at high temperature is the limit of the state of matter, which undergoes as gas $\to$ liquid $\to$ solid upon cooling.

(ii) No thermal expansion for $T \to 0$: Maxwell's equation, Equation (3.24),

$$\left(\frac{\partial V}{\partial T} \right)_P = - \left(\frac{\partial S}{\partial P} \right)_T, \quad \text{implies that,}$$

$$\lim_{T \to 0} \alpha = \lim_{T \to 0} \frac{1}{V} \left(\frac{\partial V}{\partial T} \right)_P$$

$$= \lim_{T \to 0} \frac{-1}{V} = \left(\frac{\partial S}{\partial P} \right)_T = 0$$

The last step shows that any derivative of a constant is zero.

(iii) The absolute $T = 0$ (zero point) is unattainable: It is required to reach lower temperatures by performing adiabatic and isothermal transformations, subsequently.

Using a gas as a working substance, cooling can be achieved by the lined method through a sequence of isothermal and adiabatic transformations.

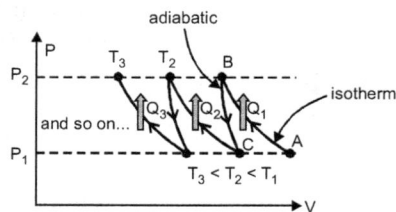

FIGURE 3.10 Lined method of cooling.

(iv) A → B isothermal compression: Work is performed on the gas and an amount of heat $Q_1 < 0$ is transferred from the matter to be cooled (characterized by a low temperature T_1) to the reservoir (having a higher temperature) in a reversible process. The entropy of the substance being cooled diminishes consequently into

$$\Delta S_1 = \frac{Q_1}{T_1}$$

(v) B → C adiabatic expansion: The gas cools by performing work. The entropy remains however (with $\delta Q = 0$) constant.

Note that all entropy curves converge to $S\ (T \to 0) \to 0$. Hence, the process becomes progressively ineffective and then infinite number of lined iterations would be needed to reach the limit $T \to 0$.

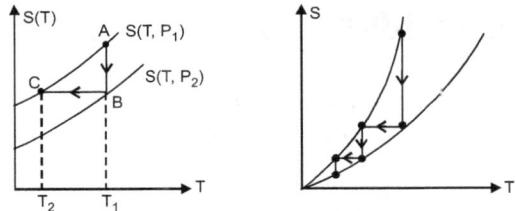

FIGURE 3.11 Isothermal compression and adiabatic expansion.

REVIEW QUESTIONS

1. What is entropy? What are its characteristics? How is it related to second law of thermodynamics?

2. Differentiate between a reversible and an irreversible process.

3. What is irreversibility? What are its sources? Differentiate between internal and external irreversibilities with examples.

4. Prove that entropy is a thermodynamic property.

5. Derive an equation of entropy balance for a closed system.

6. Derive an expression for an ideal gas undergoing isentropic process in a closed system.

7. Derive Bernoulli's equation for an incompressible flow of fluid inside a pipe.

8. Derive an equation of entropy balance for an open system.

9. Define the following:
   a. Entropy of an ideal gas.
   b. Maxwell's equation.
   c. Euler's cyclic chain rule.

10. What is the difference between *T-s* diagram and *h-s* diagram?

11. Explain the third law of thermodynamics

12. Write notes on:

    a. Isothermal compression

    b. Adiabatic expansion

NUMERICAL EXERCISES

1. A heat engine working on Carnot cycle absorbs heat from three reservoirs at 1000 K, 800K, and 600K. The engine works at 10 kW of network and rejects 400 kJ/min of heat to a heat sink at 300K. If heat supplied by the reservoir at 1000 K is 60% of the heat supplied by the reservoir at 600K, calculate the quantity of heat absorbed from each reservoir
$$\textbf{(312.5 kJ/min, 500 kJ/min, 187.5 kJ/min)}$$

2. 0.25 kg/s of water is heated from 30°C to 60°C by hot gases that enter at 180°C and leave at 80°C. Calculate the mass flow rate of gases where $C_p = 1.08$kJ/kg·K. Find the entropy change of water and hot gases. Take specific heat of water as 4.186 kJ/kg·K. Also find the increase of unavailable energy if the ambient temperature is 27°C.
$$\textbf{(0.29 kg/s, 0.0988 kJ, 0.078 kJ, 23.436 kJ)}$$

3. 10 m^3 of air at 175°C and 5 bar is expanded to a pressure of 1 bar while temperature is 30°C. Calculate the enthalpy change for the process.
$$\textbf{(3.445 kJ)}$$

4. 5 kg of ice at −10°C is kept in atmosphere which is at 30°C. Calculate the change in entropy of the universe where it mails and comes into thermal equilibrium with atmosphere. Take latent heat of fusion as 335 kJ/kg and specific heat of ice as half that of water. $\textbf{(0.652 kJ)}$

5. Using an engine of 30% efficiency to drive a refrigerator having a COP of 5, which is the heat input into the heat engine for each MJ removed from the cold body by refrigerator. If the system is used as a heat pump, how many MJ of heat would be available for heating for each MJ of heat input to the engine. $\textbf{(0.67 MJ, 1.87 MJ)}$

4

ENERGY ANALYSIS OF SYSTEMS

4.1 INTRODUCTION

The second law of thermodynamics is a very powerful tool for the analysis and the optimization of complex systems. In this chapter, a general introduction to exergy (availability) is given and explained as the maximum useful work that can be drawn from a system enclosed in a specific environment. Reversible work is defined as the maximum work that can be drawn from a process of a system between two specific states. The principle of irreversibility (destruction of energy or the lost work) is presented as a function of irreversibility. Second-law efficiency is defined. Energy balance is developed for control volume and closed system. The datum term condition of the dead state used in the calculations of the exergy or the available energy of a system is taken as the ambient conditions of the pressure P_0 and the temperature T_0, when a system is in mechanical and thermal equilibrium.

4.2 DEFINITION OF EXERGY

The word "exergy" was first used by Rant in 1956, and related functions had been presented by Keenan and Gibbs. Exergy of a system can be defined as the maximum theoretical work that can be obtained from a specific amount of energy. The term "exergy" (at a given time) can be defined as the maximum useful work that can be obtained from a system at equilibrium with the surrounding. Both energy and exergy are system properties. The more complete definition for the exergy is the work that can be obtained from the system which works ideally (under ideal conditions, reversible process) and the environment is the heat reservoir. Availability is another term used to define the exergy of a system especially involved with the heat exchange. Energy is the maximum work obtained from a system undergoing heat exchange process maintained at equilibrium with the environment.

Exergy of a system is useful to check the possibility of energy harvesting from the system, for example, from exhaust temperature of the power plants. The combination of the first and the second laws of thermodynamics is used for exergy computation. Mathematically, it can be expressed as follows:

$$\sum_{k \neq 0}\left(1 - \frac{T_0}{T_K}\right)Qk - T_0 S_{gen} = 0 \ldots \tag{4.1}$$

4.3   EXERGY ANALYSIS OF A SYSTEM

Energy contents of a system cannot be totally converted into useful work. Some portion of the system's energy will have to be rejected as waste energy. Hence, it is necessary to have a system property that enables to control the useful work potential of a specific energy of the system.

The characteristics of exergy can be stated as follows:

1. Exergy defines the maximum work that can be transferred by a system at a given time, t. For a given state of a system, there is only a single value of the maximum work, irrespective of the path taken by the system to reach its present state at time t. Exergy is a system property and it is independent on the path of the process. It is exactly like p, T, v, h, and u of the system. However, exergy depends on the state of the system and its environment.

2. Exergy of a system is zero when the system is at equilibrium with the surrounding. The exergy of a system is always positive. Negative exergy has no meaning.

3. As shown in Figure 4.1, as the system changes its state from an initial to an intermediate state, its exergy also changes. However, only some percentage of this change can be transformed into work. Regardless of the process and method, the useful work will always be less than the change in the exergy of the system.

4. System's exergy increases with its difference from its environment. Exergy is destroyed whenever energy drops its quality.

5. The surroundings are assumed to be unaffected by the system changes. Surroundings are maintained at a constant state of pressure P_0 and

temperature T_0. In order to estimate a system's exergy, the state of surroundings is taken as the reference state.

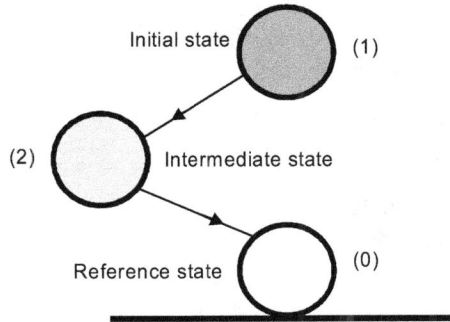

FIGURE 4.1 The change in exergy between two states.

4.3.1 Exergy Analysis of a Closed System

For the analysis of a system's exergy, the first step is to specify the initial state that makes it non-variable. Maximum work can be obtained from a process occurring between two specific states reversibly. The irreversibilities are ignored for the determination of work potential. A dead state is the state of a system that must be achieved at the end of the process to obtain the maximum work.

Exergy cannot be created, but can be destroyed.

Exergy balance of a system

(Total energy entering) – (Total exergy leaving) – (Total exergy destroyed) = (Change in the total exergy of the system)

Figure 4.2 shows exergy balance graphically. For more convenience, the exergy balance can be expressed as follows:

$$X_{in} - X_{out} - X_{destroyed} = X_{system} \qquad (4.2)$$

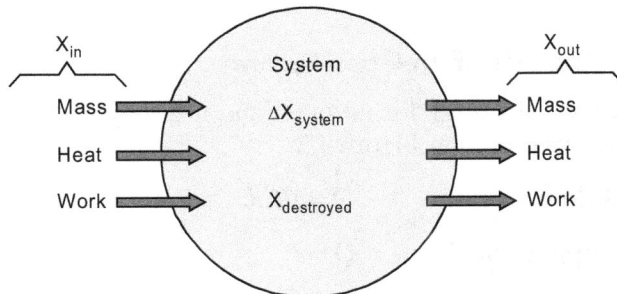

FIGURE 4.2 Exergy balance.

Equation (4.2) shows the exergy balance. The change in a system's exergy is equal to the difference between the system's exergy that is transferred through its boundary (input and output) and the destroyed exergy within the boundaries due to the system's irreversibility. The exergy is transferred either by mass, work, or heat. Therefore, exergy balance can be expressed as follows:

$$\underbrace{X_{in} - X_{out}}_{\text{Net energy transfer by heat, work and mass}} - \underbrace{X_{destroyed}}_{\text{Exergy destruction}} = \underbrace{\Delta X_{system}}_{\text{Change in exergy}} \quad (\text{KJ}) \qquad (4.3)$$

The exergy transfer rates are in terms of mass, work, and heat and can be expressed as follows:

$$\left.\begin{aligned} \dot{X}_{heat} &= \left(1 - \frac{T_0}{T}\right)\dot{Q} \\ \dot{X}_{work} &= \dot{W}_{useful} \\ \dot{X}_{mass} &= \dot{W}_{\varphi} \end{aligned}\right\} \qquad (4.4)$$

The term "$X_{destroyed}$" is cancelled out from the reversible process. To obtain the destroyed exergy, the term "S_{gen}," that is entropy generated, must be obtained first. The destroyed exergy can be calculated as follows:

$$X_{destroyed} = T_0 \ S_{gen} \qquad (4.5)$$

For the closed system, there is no mass transfer. The exergy balance for a closed system will be,

$$X_{heat} - X_{work} - X_{destroyed} = X_{system}$$

And,

$$\sum\left(1 - \frac{T_0}{T_k}\right)Q_k - [W - P_0(V_2 - V_1)] - T_0 S_{gen} = X_2 - X_1 \qquad (4.6)$$

4.3.2 Exergy Analysis of an Open System

The balance of exergy for an open and a steady state system with constant volume is presented in Figure 4.3.

Reversible processes: $\qquad Q = \int T.ds \qquad (4.7)$

Heat is supplied at T_0: $\qquad Q = T_0 \cdot (S_0 - S) \qquad (4.8)$

$$W_{rev} = Ex_{matter} = (H - H_0) - T_0 \cdot (S - S_0) \qquad (4.9)$$

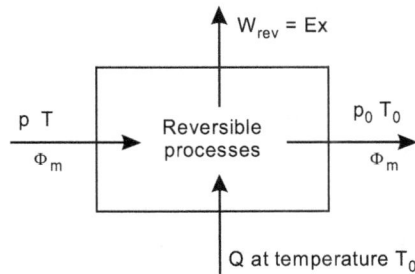

FIGURE 4.3 Open steady state system.

And for a flow of mass:

$$\phi_{w_{rev.}} = \phi_{Ex_{flow}} = \phi_m \left[(h - h_0) - T_0 (s - s_0) \right] \qquad (4.10)$$

4.4 EXERGY LOSS OF A SYSTEM

In all engineering processes, the following factors may cause a loss of exergy of the system:

a. Electrical resistance

b. Inelastic deformations

c. Viscous fluid flow

d. Friction: solid-to-solid, solid-to-fluid, fluid-to-fluid

e. Shockwaves

f. Damping of vibrating systems

g. Fluid behavior at sudden expansion

h. Fluid flow through a valve or throttling process

i. Heat transfer by temperature difference, "thermal friction"

j. Sudden occurrence of chemical reactions

k. Mixing of liquids or gases having different chemical compositions

l. Process of osmosis

m. Phase existing in another face

n. Mixing of fluids at different pressures and temperatures

The exergy loss of a system is zero if it undergoes a reversible process.

Since there is no loss in a system's exergy for reversible process, the system's exergy at time t shows the maximum possible work that can be obtained till the system comes into common equilibrium with the environment. The reverse is also true. For a reversible process without the need of any additional energy, the amount of work extracted from the system can be supplied back, and hence the initial state conditions of the system and the environment can be restored. Consequently, there will be no noticeable change in the system or in its environment. System irreversibility can be measured by the amount of exergy loss due to irreversible process. However, exergy loss is zero for a reversible process. The irreversibility can be classified into two categories: external and internal.

4.4.1 External Irreversibility

The system's irreversibility that takes place due to the system reaction or system interaction with the environment is called external irreversibility. Examples are friction, pressure changes, heat transfer due to temperature difference, etc.

4.4.2 Internal Irreversibility

This irreversibility takes place when the properties of the system are distributed heterogeneously. Local temperature, pressure, and velocity gradients within the system lead to sudden processes, which lead to exergy loss. For example, the presence of viscosity of fluid flow inside a channel results in different particle velocities. The velocity differences develop a rubbing action between the particles producing heat. The presence of heat transfer results in exergy loss which leads to irreversibility. This is called internal irreversibility because it occurs inside the system.

4.4.3 Exergy Loss of an Open, Steady State, Constant Volume System

Exergy balance of an irreversible open steady state system is shown in Figure 4.4.

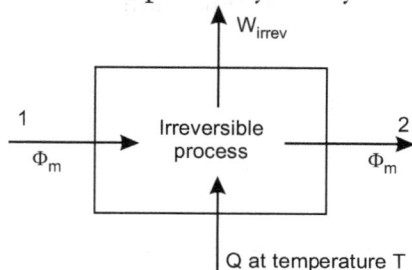

FIGURE 4.4 Irreversible open steady state system.

$$Ex_{loss} = Ex_{in} - Ex_{out} \tag{4.11}$$

For the system under consideration:

$$Ex_{loss} = Ex_1 - Ex_2 + \int_1^2 \left(1 - \frac{T_0}{T}\right) \cdot dQ - W_{irrev}$$

$$\therefore \quad Ex_{loss} = Ex_1 - Ex_2 + \int_1^2 \left(1 - \frac{T_0}{T}\right) \cdot dQ - W_{irrev}$$

But
$$Ex_1 = (H_1 - H_2) - T_0(S_1 - S_2)$$
$$Ex_2 = (H_2 - H_0) - T_0(S_2 - S_0)$$
or $\quad Ex_2 = (H_1 - H_2) - T_0(S_1 - S_2) \tag{4.12}$

$$\therefore \quad Ex_{loss} = (H_1 - H_2) - T_0(S_1 - S_2) + Q - T_0 \int_1^2 \left(\frac{dQ}{T}\right) - W_{irrev.} \tag{4.13}$$

From the first law, the energy balance of an open steady state system is:

$$Q = (H_2 - H_1) + W_{irrev} \rightarrow Q + (H_1 - H_2) - W_{irrev.} = 0$$

The exergy balance becomes:

$$Ex_{loss} = T_0 \left((S_2 - S_1) - \int_1^2 \left(\frac{dQ}{T}\right)\right) \tag{4.14}$$

The entropy balances is:

$$(S_2 - S_1) = \int_1^2 \left(\frac{dQ}{T}\right) + \Delta S_{irrev} \rightarrow \Delta S_{irrev}$$

$$= (S_2 - S_1) - \int_1^2 \left(\frac{dQ}{T}\right)$$

By combining these two equations,

$$Ex_{loss} = T_0 \cdot \Delta S_{irrev} \tag{4.15}$$

4.5 EXERGY EQUATION

Transferring system's exergy can be defined as the system that allows the energy to pass through the boundaries provided that some of this energy can be transformed into a useful work. This definition indicates that exergy and energy are interrelated. In other words, exergy transfer cannot take place without energy transfer. The rate of exergy change for a system can be presented by Equation (4.16).

$$\left\{ \begin{array}{l} \text{Time rate of exergy accumulation} \\ \text{within a system at an instant of time } t \end{array} \right\} = \left\{ \begin{array}{l} \text{The net rate of exergy transport} \\ \text{into the system at time } t \end{array} \right\}$$

$$+ \left\{ \begin{array}{l} \text{The time rate of irreversibility} \\ \text{production at an instant of time } t \end{array} \right\}$$

(4.16)

The energy transfer between two systems can be accomplished in three different modes. Similarly, the exergy transfer within a system can take place in three different modes, that is, the exergy transfer by convection, heat, and work.

4.5.1 Transfer of Exergy by Work

The exergy transfer by a property difference other than the temperature is called transfer of exergy by work. For example, in the case of work transfer accomplished by the boundaries motion, the useful work can be obtained by removing the portion of the work done by the environment. The exergy rate due to work can be calculated as follows:

$$\dot{x}_w = \dot{W} - P_0 \frac{dV}{dt} \tag{4.17}$$

Following example gives the procedure for calculation of the exergy loss of a system.

EXAMPLE 4.1

An insulated rigid tank, as shown in Figure 4.5, holds 1 kg of pressurized air at 1 bar and maintained at 27°C. A motor runs a fan to perform an external work so that the final temperature of the air is 527°C. This results in exergy increasing to 100 kJ. Find the losses in exergy and explain the reason for the same.

Solution: The energy equation for the adiabatic tank is as follows:

$$U_2 - U_1 = -W_{12},$$

For ideal gas,

$$U_2 - U_1 = mc_v \left(T_2 - T_1 \right) = W_{12} = -358 \text{ kJ}.$$

Rigid tank prevents volume change and thus no work transfer occurs through the boundaries. Exergy increment of the system will be 358 kJ for the reversible process. However, the change of exergy is given as Df = 100 kJ and the exergy transferred by the work is Dx_w = –358 kJ.

The main reason of additional exergy losses is due to the friction developed between the following:

a. The fluid and the fan blades

b. The fluid and the surface of the tank

c. The fluid layers

d. The shaft housing and its surface

e. The bearing of the housing

Exergy cannot be conserved and that contradicts the principle of energy conservation. The friction is a critical factor for exergy loss during work transfer between two systems.

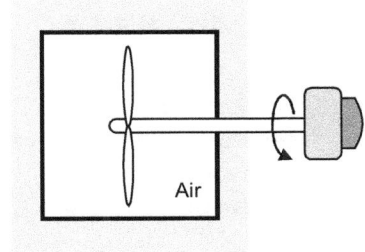

FIGURE 4.5 Adiabatic tank.

4.5.2 Transfer of Exergy by Heat

Temperature difference causes exergy transfer due to heat flow across the boundaries of two systems. Figure 4.6 shows a reversible power cycle, where heat is rejected to the environment at T_0. The maximum theoretical work from a reversible power cycle can be calculated as follows:

$$dW_{rev} = \left(1 - \frac{T_C}{T_H}\right) . dQ$$

With $T_H = T$ and $T_C = T_0$:

$$dW_{rev} = \left(1 - \frac{T_0}{T_H}\right) . dQ \text{ and } dW_{rev} = dE_{XQ}$$

$$\therefore \quad dX_{EQ} = \left(1 - \frac{T_0}{T}\right) . dQ$$

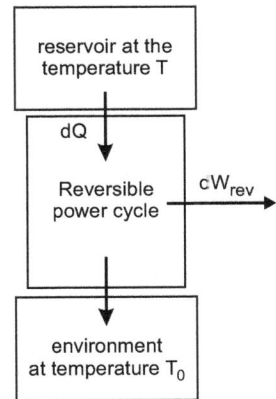

FIGURE 4.6 Reversible power cycle.

Exergy of heat: $\quad dX_{EQ} = \left(1 - \frac{T_0}{T}\right) . dQ \qquad (4.18)$

Heat is transferred to the system at varying temperature:

$$E_{XQ} = \int_1^2 dE_{XQ} = \int_1^2 \left(1 - \frac{T_0}{T}\right) dQ$$

Or
$$E_{XQ} = \left(1 - \frac{T_0}{T}\right) \cdot Q,$$
(4.19)

where $\overline{T}$ is *thermodynamic* equivalent temperature of heat transfer to the cycle.

EXAMPLE 4.2

Figure 4.7 shows a reversible heat engine working between two thermal reservoirs at 227°C. The heat engine supplies work to a refrigeration system (reversible). The refrigeration system is used to cool the space at −50°C by rejecting the heat at 27°C to an environment. Calculate the ratio Q_1/Q_2.

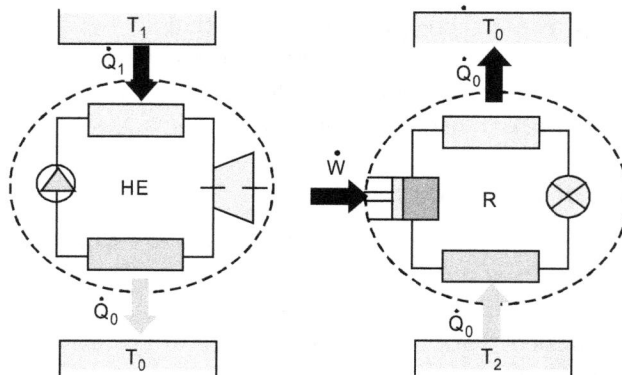

FIGURE 4.7 A reversible heat engine.

Solution: For the reversible heat engine, the output work is calculated as follows:

$$\dot{W}_{he} = \dot{Q}_1 \left(1 - \frac{T_0}{T_1}\right)$$

The work consumed by the refrigeration system is calculated as follows:

$$\dot{W}_{he} = \dot{Q}_1 \left(1 - \frac{T_0}{T_1}\right)$$

Since, $\dot{W}_{he} = \dot{W}_r$

The heat ratio becomes:

$$\frac{\dot{Q}_1}{\dot{Q}_2} = \frac{\dfrac{T_0}{T_2} - 1}{1 - \dfrac{T_0}{T_1}}$$

By substituting the temperature values as $T_1 = 500$ K, $T_0 = 300$ K and $T_2 = 223$ K,

$$\frac{\dot{Q}_1}{\dot{Q}_2} = 0.862$$

EXAMPLE 4.3

Figure 4.8 shows a house heating system. It consists of a reversible heat pump. In order to maintain the temperature inside the house at 22°C, the required heat (2°C outside temperature) is 12×10^4 kJ/h. The work done by the compressor is 10 kW.

a. Calculate the number of hours for one day of pump work.

b. The house heating could be accomplished using electric heaters. Find the useful energy that can be conserved using the heat pump. The determined useful energy is required on a daily basis.

Solution: The inside and the outside temperatures are $T = 295$ K and $T_0 = 275$ K, respectively.

$$\dot{Q}_1 = \frac{\dot{W}_{he}}{1 - \dfrac{T_0}{T_1}}$$

$$\dot{Q}_1 = \frac{10}{1 - \dfrac{275}{295}} = 147.5 \text{ kW}$$

The heat loss of the house is:

$$\dot{Q}_h = \frac{120000}{3600} = 33.3 \text{ kW}$$

a. Number of hours required:

$$t_{hp} = \frac{\dot{Q}_h}{\dot{Q}} \times 24$$

$$t_{hp} = \frac{33.3}{147.5} \times 24$$

$$t_{hp} = 5.418 \text{ hours}$$

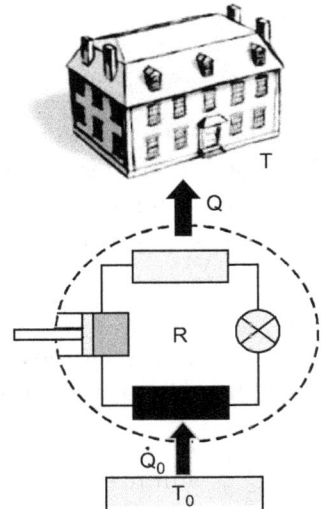

FIGURE 4.8 Heating of a house.

b. The electrical energy (using the heat pump) can be calculated as follows:

$$E_{hp} = (\dot{W}_t)_{hp}$$
$$E_{hp} = 10 \times 5.418 = 54.18 \text{ kWh}$$

The electrical energy (using the heaters) is $E_h = 33.3 \times 24 = 799.2$ kWh

The difference between the electrical energy consumed by the heaters and the heat pumps (799.2 − 54.18 = 745.02 kWh) is the useful energy that can be gained.

4.5.3 Transfer of Exergy by Convection

Fluid flow that occurs in and out of a system causes exergy transfer by convection. The net exergy accumulation rate due to transfer of energies by heat, work, and conviction are as follows:

{The net rate of exergy transfer by work at an instant of time, t} = $\dot{X}_W$

{The net rate of exergy transfer by heat at an instant of time, t} = $\dot{X}_Q$

{The net rate of exergy accumulation by convection at an instant of time, t}

$$= \sum_i \dot{m}_i \psi_i - \sum_e \dot{m}_e \psi_e$$

The net rate of exergy supplied into the system will be

$$\left\{ \begin{array}{l} \text{The net rate of exergy accumulation by convection} \\ \text{in the system at an instant of time } t \end{array} \right\} =$$

$$\sum_i \dot{m}_i \psi_i - \sum_e \dot{m}_e \psi_e + \dot{X}_Q - \dot{X}_W \qquad (4.20)$$

As a convention, one can assume that the symbol $\dot{I}$ represents the irreversibility rate at time t. It is known that the reversible process shows no irreversibility rate ($\dot{I} = 0$). However, the inequality $\dot{I} \geq 0$ is true for the irreversible process. Based on that, Equation (4.20) can be rearranged as follows:

$$\phi = \sum_i \dot{m}_i \psi_i - \sum_e \dot{m}_e \psi_e + \dot{X}_Q - \dot{X}_W - \dot{I} \qquad (4.21)$$

where f is the exergy accumulation rate in the system at time t and rate of the destroyed exergy $\dot{X}_{destroyed}$ is the irreversibility rate $\dot{I}$. This means,

$$\dot{X}_{destroyed} = \dot{I} ... \qquad (4.22)$$

For better understanding of Equation (4.20), all the terms $\dot{X}_Q, \Psi_i$, and $\dot{X}_w$ must be found out.

REVIEW QUESTIONS

1. Define energy and explain its relevance.

2. With the help of switchable diagrams, explain the exergy analysis of the following:
   a. Closed system
   b. Open system

3. List out the various factors that cause exergy loss in a system.

4. Develop a suitable exergy equation for a system and explain the following:
   a. Transfer of exergy by work
   b. Transfer of exergy by heat
   c. Transfer of exergy by convection

5. With a suitable example, prove that heating of a house with the help of a heat pump is more efficient than heating by electric heaters.

NUMERICAL EXERCISES

1. Calculate the physical exergy of a perfect gas (air, mass (m) = 3 kg, isobaric specific heat C_p = 1.0005 kJ/kg·K, and gas constant R = 0.287 kJ/kg·K) for a state defined by p_1 = 3 bar and T_1 = 398 K. The environmental parameters are p_0 = 1 bar and T_0 = 298 K.   **(323.4 kJ)**

2. A flat plate collector heats water from 305 K to 335 K. Calculate the exergy efficiency of the collector if its efficiency is 48% and ambient temperature is 293 K.   **(8%)**

3. Air is heated in a central receiver solar power plant from 693 K to 1073 K. Calculate the exergy efficiency if ambient temperature is 303 K and collector efficiency is 70%.   **(42.5%)**

5

THERMODYNAMIC RELATIONS

5.1   TYPES OF THERMODYNAMIC PROPERTIES

The condition of thermodynamic system can be defined by its character-istics called properties. The properties can be essential, derived, or meas-ured. Relationships can be developed to correlate the changes in the derived properties with reference to the measured properties. Some of the measured properties are V, P, T, c_p, c_v, and composition. Small letters are used to indi-cate the specific quantities, for example, v denotes the specific volume. Using first and second laws of thermodynamics, the internal energy (u) and entropy (s) are considered as the fundamental derived properties. As per first law, the energy is conserved and as per second law, entropy of the universe continues to increase. The derived properties are used for promoting the energy bal-ance of systems where internal energy and other properties are frequently combined.

The mass flow in the open systems that crosses the boundary between the system and the surroundings provides two types of energy: flow work (pv) and internal energy (u).

An enthalpy (h) can be defined as follows:

$$h = u + pv \tag{5.1}$$

Or total enthalpy can be defined as follows:

$$H = U + PV \tag{5.2}$$

For an open system, flow work is included in the enthalpy along with internal energy (U).

In an isothermal system, heat can be extracted from the surroundings for free. For any environment having constant temperature T, *Helmholtz free energy* (A) is very suitable.

$$A = U - TS \tag{5.3}$$

A denotes work if the system is created from nothing. The heat extracted from the surroundings is $T\Delta S = T\,(S_f - S_i) = TS_f$, where S_f is the final entropy of the system, while S_i represents its zero-initial entropy. If a system is terminated with an initial entropy S_i, A denotes the recovered work, as some heat needs to be dumped (TS_i) into the environment for eradicating the entropy of the system. Equation (5.3) covers all work including the work performed by the surroundings. It is more convenient to use *Gibbs free energy* in an isobaric and isothermal environment.

$$G = U - TS + PV \tag{5.4}$$

Gibbs free energy can be defined as the work that is needed for generating a system out of nothing within an environment having constant pressure (P) and temperature (T).

The changes in the state of a system are more practical for which the changes in A and G should be considered. Changing A and keeping the temperature constant leads to:

$$\Delta A = \Delta U - T\Delta S = W + Q + T\Delta S \tag{5.5}$$

Q denotes the heat added and W represents the system's work. For reversible process, Q is equal to $T\Delta S$ and the change in A is equivalent to the work done by the system. However, for irreversible process, Q is lower than $T\Delta S$ and ΔA is lower than W. The change of A is less than the work done by the system.

At constant temperature (T) and pressure (P), the change in G for the environment is calculated as follows:

$$\Delta G = \Delta U - T\Delta S + P\Delta V = W + Q - T\Delta S + P\Delta V \tag{5.6}$$

For any process,

$$Q - T\Delta S \leq 0 \ \text{(similar signs for processes with reversibility)} \tag{5.7}$$

The work (W) includes the work done to the environment ($-P\Delta V$) and any "other" work achieved on the system.

$$W = -P\Delta V W_{other} \tag{5.8}$$

Substituting Equations (5.7) and (5.8) into Equation (5.6), the following was obtained.

$$\therefore \Delta G \leq W_{other} \ \text{at constant } T, P \tag{5.9}$$

5.2   EQUATIONS OF STATE

For the calculation of entropy, enthalpy, and energy of a matter, a correlation among volume, pressure, and temperature is required. In addition to the graphical and tabular configuration of the p-v-T diagram, there are analytical expressions termed as equations of state, which represents another approach of the p-v-T correlation. The state equations are used for calculating h, u, s, and other thermodynamic properties.

5.2.1 Virial Equation of State

The Virial equation of state can be derived from the static mechanics by relating the p-v-T behavior of any gas to the forces between its molecules. The Virial equation of state defines the term $\dfrac{PV}{RT}$ as a power-series in the inversion of molar volume ($\overline{v}$).

$$Z = \frac{P\overline{v}}{RT} = 1 + \frac{B(T)}{\overline{v}} + \frac{C(T)}{\overline{v}^2} + \frac{D(T)}{\overline{v}^3} + \ldots, \tag{5.10}$$

where B, C, and D are Virial coefficients and acts as the functions of temperature.

Simplifying Equation (5.10), the following was obtained:

$$\frac{P\overline{v}}{RT} = 1 + \frac{B(T)}{\overline{v}} \tag{5.11}$$

$B(T)$ can be calculated as follows:

$$B(T) = \frac{\overline{R}T_c}{P_c}(\beta_0 + \omega\beta_1), \tag{5.12}$$

where $$\beta_0 = 0.083 - \frac{0.422}{T_R^{1.6}}, \quad \beta_1 = 0.139 - \frac{0.172}{T_R^{4.2}}$$

ω denotes the Pitzer acentric factor. This parameter describes the polarity and the geometry of a molecule. A program written by T. K. Nguyen named "comp4.exe" could be used to find the Pitzer acentric factor for over 1000 compounds.

Comp4.exe can be found in the CHE302 course Distribution Folder. In some cases, where the molecular interactions do not exist, all Virial coefficients are equal to zero. Since the coefficients are equal to zero, Equation (5.10) becomes,

$$Z = \frac{P\overline{v}}{\overline{R}T} = 1 \tag{5.13}$$

Equation (5.13) is the ideal gas equation of state.

5.2.2 Van der Waals Equation of State

Van der Waals equation of state could be used to show the assessment of thermodynamic properties. It looks easier than Virial equation of state.

Van der Waals equal of state is as follows:

$$P = \frac{\overline{R}T}{\overline{v} - b} = \frac{a}{\overline{v}^2}, \tag{5.14}$$

where b is a constant related with the occupied finite volume by the molecules, and $\frac{a}{\overline{v}^2}$ is related with the attractive forces between molecules.

a and b can be calculated using critical properties because the critical isotherm has an inflection point as shown in Figure 5.1. In the case of critical point, it will be as follows:

$$\left(\frac{\partial P}{\partial \overline{v}}\right)_{T_c} = \left(\frac{\partial^2 P}{\partial \overline{v}^2}\right)_{T_c} = 0 \tag{5.15}$$

The isotherm that passes through the critical point is as follows:

$$P = \frac{T_c \overline{R}}{\overline{v} - b} - \frac{a}{\overline{v}^2} \tag{5.16}$$

The first and second derivatives of P with $\overline{v}$ give the following:

$$\left(\frac{\partial P}{\partial \overline{v}}\right)_{T_c} = -\frac{\overline{R}}{(\overline{v}_c - b)^2} + \frac{2a}{\overline{v}_c^3} = 0 \tag{5.17}$$

and

$$\left(\frac{\partial^2 P}{\partial \overline{v}^2}\right)_{T_c} = \frac{2\overline{R}T_c}{(\overline{v}_c - b)^3} - \frac{6a}{\overline{v}_c^4} = 0 \tag{5.18}$$

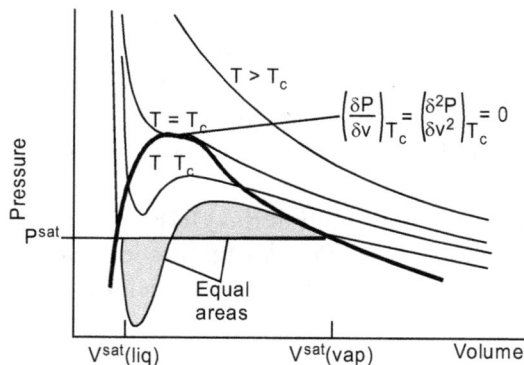

FIGURE 5.1 Van der Waals equation for isotherms.

Multiplying Equation (5.17) by 2 and Equation (5.18) by $(\bar{v}_c - b)$ and adding together to get the following:

$$\frac{4a}{\bar{v}_c^3} - \frac{6a}{\bar{v}_c^4}(\bar{v}_c - b) = 0 \tag{5.19}$$

$$4a\bar{v}_c - 6a\bar{v}_c + 6ab = 0 \tag{5.20}$$

$$\bar{v}_c = 3b$$

If $b = \dfrac{\bar{v}_c}{3}$ is substituted into Equation (5.18), the following was obtained:

$$a = \frac{9}{8}\bar{v}_c \bar{R} T_c$$

In the case of critical point, it will be as follows:

$$P_c = \frac{\bar{R}T_c}{\bar{v}_c - b} - \frac{a}{\bar{v}_c^2} \tag{5.21}$$

Equation (5.21) can be used for solving a and b in terms of the critical pressure and critical temperature. When $a = \dfrac{9}{8}\bar{v}_c \bar{R} T_c$ and $b = \dfrac{\bar{v}_c}{3}$ are substituted into Equation (5.21), the following was obtained:

$$P_c = \frac{3\bar{R}T_c}{2\bar{v}_c} - \frac{9\bar{v}_c \bar{R} T_c}{8\bar{v}_c} = \frac{3\bar{R}T_c}{2\bar{v}_c}\left(\frac{3}{2} - \frac{9}{8}\right) = \frac{3\bar{R}T_c}{8\bar{v}_c}$$

$$\therefore \bar{v}_c = \frac{3\bar{R}T_c}{8P_c} \tag{5.22}$$

and

$$a = \frac{9}{8}\bar{v}_c \bar{R} T_c = \frac{27}{64}\frac{(\bar{R}T_c)^2}{P_c}$$

5.3 MAXWELL'S RELATIONS

5.3.1 Energy and Entropy

Entropy $S(V, U, N)$ and energy $U(V, S, N)$ can give complete information about thermodynamics.

The important equation of thermodynamics for energy is as follows:

$$dU = T\,dS - p\,dV + \mu dN$$

This can be used to find the chemical potential, temperature, and pressure at equilibrium conditions. The partial derivatives are as follows:

$$\left(\frac{\partial U}{\partial S}\right)_{V,N} = T \tag{5.23}$$

$$\left(\frac{\partial U}{\partial V}\right)_{S,N} = -P \tag{5.24}$$

And
$$\left(\frac{\partial U}{\partial N}\right)_{S,N} = \mu \tag{5.25}$$

Important equation of thermodynamics for entropy is as follows:

$$dS = \frac{dU}{T} + \frac{P}{T}dV - \frac{\mu}{T}dN \tag{5.26}$$

The partial derivatives will be as follows:

$$\left(\frac{\partial S}{\partial U}\right)_{V,N} = \frac{1}{T} \tag{5.27}$$

$$\left(\frac{\partial S}{\partial V}\right)_{U,N} = \frac{P}{T} \tag{5.28}$$

And
$$\left(\frac{\partial S}{\partial N}\right)_{U,N} = \frac{\mu}{T} \tag{5.29}$$

Maxwell's relations can be found by taking the second partial derivatives of these qualities. Such relationships can be used for connecting the partial derivatives, of properties which can be measured with those which cannot be measured.

Starting from:

$$\left(\frac{\partial U}{\partial S}\right)_{V,N} = T \text{ and } \left(\frac{\partial U}{\partial V}\right)_{S,N} = -P,$$

Determine:

$$\frac{\partial^2 U}{\partial V\,\partial S} = \left(\frac{\partial T}{\partial V}\right)_{S,N} \quad \text{and} \quad \frac{\partial^2 U}{\partial S\,\partial V} = -\left(\frac{\partial P}{\partial S}\right)_{V,N}$$

Under appropriate conditions,

$$\frac{\partial^2 U}{\partial V\,\partial S} = \left(\frac{\partial^2 U}{\partial S\,\partial V}\right) \tag{5.30}$$

And
$$\left(\frac{\partial T}{\partial V}\right)_{N,S} = -\left(\frac{\partial P}{\partial S}\right)_{N,V} \tag{5.31}$$

Equation (5.31) is known as the Maxwell's relation. By performing the other second partial derivatives, another two Maxwell's relations can be found.

In terms of energy,

$$\left(\frac{\partial T}{\partial V}\right)_{N,S} = \left(\frac{\partial \mu}{\partial S}\right)_{N,V} \text{ and } \left(\frac{\partial P}{\partial N}\right)_{V,S} = \left(\frac{\partial \mu}{\partial V}\right)_{N,S}$$

In terms of entropy,

$$dS = \frac{dU}{T} + \frac{P}{T}dV - \frac{\mu}{T}dN$$

And the results are as follows:

$$\left(\frac{\partial S}{\partial U}\right)_{N,V} = \frac{1}{T}, \quad \left(\frac{\partial S}{\partial V}\right)_{N,U} = \frac{P}{T}, \quad \text{and} \quad \left(\frac{\partial S}{\partial N}\right)_{N,U} = -\frac{\mu}{T}$$

The Maxwell's relations are as follows:

$$\left(\frac{\partial\left(\frac{1}{T}\right)}{\partial V}\right)_{N,U} = \left(\frac{\partial\left(\frac{P}{T}\right)}{\partial U}\right)_{N,V}$$

$$\left(\frac{\partial\left(\frac{1}{T}\right)}{\partial N}\right)_{N,U} = \left(\frac{\partial\left(\frac{\mu}{T}\right)}{\partial U}\right)_{V,N}$$

And

$$\left(\frac{\partial\left(\frac{P}{T}\right)}{\partial N}\right)_{V,U} = \left(\frac{\partial\left(\frac{\mu}{T}\right)}{\partial V}\right)_{N,U}$$

5.3.2 Enthalpy H (S, p, N)

The enthalpy has been defined as $H = U + PV$. Its differential can be combined with the basic thermodynamic equation to get dH in terms of S, p_0, and N.

Now,
$$H = U + pV$$

$$\therefore \quad dH = dU + d(pV)$$

$$= dU + pdV + V\,dp \tag{5.32}$$

But
$$dU = T\,dS - p\,dV + \mu\,dN \tag{5.33}$$

$$dH = T\,dS - p\,dV + \mu\,dN + p\,dV + V\,dp \tag{5.34}$$

or
$$dH = T\ dS + V\ dp + \mu\ dN \tag{5.35}$$

$H = H\ (S, p, N)$, where S, p, and N are the natural properties of H. Differentiating partially,

$$\left(\frac{\partial H}{\partial S}\right)_{N,P} = T, \quad \left(\frac{\partial H}{\partial P}\right)_{N,S} = V, \quad \text{and} \quad \left(\frac{\partial H}{\partial N}\right)_{P,S} = \mu$$

The Maxwell's relations are as follows:

$$\left(\frac{\partial T}{\partial P}\right)_{N,S} = \left(\frac{\partial V}{\partial S}\right)_{N,P}$$

$$\left(\frac{\partial T}{\partial N}\right)_{P,S} = \left(\frac{\partial \mu}{\partial S}\right)_{N,P}$$

and
$$\left(\frac{\partial V}{\partial N}\right)_{P,S} = \left(\frac{\partial \mu}{\partial P}\right)_{N,S}$$

5.3.3 Helmholtz Free Energy $F\ (T, V, N)$

To transform from U to F, replace independent variable S by its conjugate T. In a Legendre transformation, a new function is described by adding or subtracting conjugate product to replace an independent variable with its conjugate. Therefore, by subtracting TS from U, the new function F was defined.

Now,
$$F(T,\ V,\ N) = U(S,\ V,\ N) - TS \tag{5.36}$$

Differentially,
$$dF = dU - d(TS)$$
$$= dU - T\ dS - S\ dT \tag{5.37}$$

But
$$dU = T\ dS - p\ dV + \mu\ dN \tag{5.38}$$

$$\therefore\ dF = T\ dS - p\ dV + \mu\ dN - T\ dS - S\ dT$$

or
$$dF = -S\ dT - p\ dV + \mu\ dN \tag{5.39}$$

Therefore, $F = F\ (T, V, N)$ was preferred. By following the same procedures, the following can be defined:

$$\left(\frac{\partial F}{\partial T}\right)_{N,V} = -s, \left(\frac{\partial F}{\partial V}\right)_{N,T} = -P, \text{and} \left(\frac{\partial F}{\partial N}\right)_{V,T} = \mu$$

The Maxwell's relations are as follows:

$$\left(\frac{\partial S}{\partial V}\right)_{N,T} = \left(\frac{\partial P}{\partial T}\right)_{N,V}$$

$$\left(\frac{\partial S}{\partial N}\right)_{V,T} = -\left(\frac{\partial \mu}{\partial T}\right)_{N,V}$$

and
$$\left(\frac{\partial P}{\partial N}\right)_{V,T} = -\left(\frac{\partial \mu}{\partial V}\right)_{N,T}$$

5.3.4 Gibbs Free Energy G (T, N, p)

The independent variables V and S are replaced by their conjugates p and T, while transforming from U to G. This represents a twice Legendre transformation of U or one Legendre transformation of either F or H.

Now,
$$G(T, N, p) = U(S, N, V) - TS + pV \tag{5.40}$$

Differentiating,

$$dG = dU - d(TS) + d(pV) = dU - T\,dS - S\,dT + p\,dV + V\,dp,$$

But
$$dU = T\,dS - p\,dV + \mu\,dN$$

$$\therefore \quad dG = T\,dS - p\,dV + \mu\,dN - T\,dS + S\,dT + p\,dV + V\,dp$$

or
$$dG = -S\,dT + V\,dp + \mu\,dN. \tag{5.41}$$

$$\therefore \quad G = G(T, p, N) \text{ was preferred.}$$

Following the same procedures, the following can be defined:

$$\left(\frac{\partial G}{\partial T}\right)_{N,P} = -S, \left(\frac{\partial G}{\partial P}\right)_{N,T} = V, \text{ and } \left(\frac{\partial G}{\partial N}\right)_{P,T} = \mu$$

The Maxwell's relations are as follows:

$$\left(\frac{\partial S}{\partial P}\right)_{N,T} = -\left(\frac{\partial V}{\partial T}\right)_{N,P}$$

$$\left(\frac{\partial S}{\partial N}\right)_{P,T} = -\left(\frac{\partial \mu}{\partial T}\right)_{N,P}$$

and
$$\left(\frac{\partial V}{\partial N}\right)_{P,T} = \left(\frac{\partial \mu}{\partial P}\right)_{N,T}$$

5.3.5 The Grand Potential Ω (T, V, μ)

Here, the independent variables N and V are replaced by their conjugates μ and p, while transforming from U to Ω. It can be done by a double transformation of U Legendre or a single transformation of F. For engineering work, the ground potential is less frequently used than the other potentials.

In an open system, where particles are exchanged with the surrounding, ground potential is used.

Now
$$\Omega(T, V, \mu) = U(S, V, N) - TS - \mu N \qquad (5.42)$$

Differentiating
$$d\Omega = dU - d(TS) - d(\mu N)$$
$$= dU - S\ dT - TV\ dS - \mu\ dN - N\ d\mu,$$

But
$$dU = T\ dS - p\ dV + \mu dN$$

$$\therefore d\Omega = T\ dS - p\ dV + \mu\ dN - T\ dS - S\ dT - \mu\ dN - N\ d\mu$$

or
$$d\Omega = -pdV - S\ dT - N\ d\mu... \qquad (5.43)$$

Therefore, $\Omega = \Omega\ (T, V, \mu)$.

Differentiating again

$$\left(\frac{\partial \Omega}{\partial V}\right)_{T,\mu} = -P, \left(\frac{\partial \Omega}{\partial T}\right)_{V,\mu} = -S, \text{ and} \left(\frac{\partial \Omega}{\partial \mu}\right)_{T,V} = -N$$

The Maxwell's relations are as follows:

$$\left(\frac{\partial P}{\partial T}\right)_{V,\mu} = \left(\frac{\partial S}{\partial V}\right)_{T,\mu}$$

$$\left(\frac{\partial P}{\partial \mu}\right)_{T,V} = -\left(\frac{\partial N}{\partial V}\right)_{T,\mu}$$

and
$$\left(\frac{\partial S}{\partial \mu}\right)_{T,V} = \left(\frac{\partial N}{\partial V}\right)_{T,\mu}$$

EXAMPLE 5.1

Blender blades are churning the fluid at 97°C. The work is transferred to the fluid at a rate of 360 kJ/h. The blender is shown in Figure 5.2. The bowl of the blender is insulated circumferentially. The bottom surface of the bowl passes heat at a rate of 360 kJ/h to the surrounding at 27°C. Calculate:

A. The net energy rate supplied to the system,

B. Energy accumulation rate of the system

C. Irreversibility rate of the process.

Solution:

A. The heat transfer and work rates, respectively, are

$$\dot{Q} = 360 \text{ kJ/h}$$

$$W = 360 \text{ kJ/h.}$$

Energy balance of the system shows that

$$\dot{E} = \dot{Q} - \dot{W}, \ \dot{E}$$

$$= -360 - (-360)$$

$$= 0 \quad \textbf{Ans.}$$

No change in the system's energy and it is a steady state system.

B. The constant energy system shows no exergy.

$$\therefore \qquad \phi = 0$$

The exergy transfer by work is calculated as follows:

$$\dot{X}_w = -\dot{W}, \ \dot{X}_w = 360 \text{ kJ/h} \qquad \textbf{Ans.}$$

The transfer of exergy by heat is calculated as follows:

$$\dot{X}_Q = \dot{Q}\left(1 - \frac{300}{370}\right) = -68.108 \text{ kJ/h} \qquad \textbf{Ans.}$$

C. From the equation of exergy balance:

$$0 = -68.108 + 360 - \dot{I}$$

Irreversibility $\dot{I} = 291.89 \text{ kJ/h}$ **Ans.**

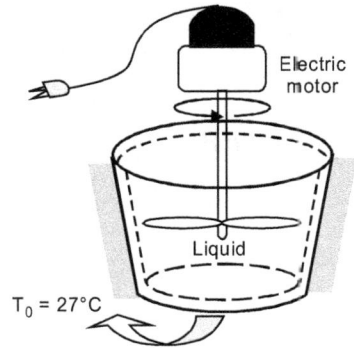

FIGURE 5.2 Blender.

$T_0 = 27°C$

5.4 FLOW EXERGY OF MATTER

The characteristics of a system used to find the flow energy of matter are as follows:

- System with only reversible processes.
- Steady state flow in an open system with constant volume.
- Heat is transferred only to and from the environment at T_0.

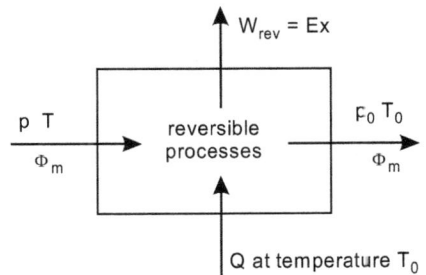

FIGURE 5.3 Matter in equilibrium.

- System maintains the equilibrium between the matter and the environment, as shown in Figure 5.3.

5.5 EXERGY EFFICIENCIES

Energy efficiency can be given by

$$\eta_{ex} = \frac{\sum Ex_{product}}{\sum Ex_{source}} = \frac{\sum Ex_{source} - \sum Ex_{loss}}{\sum Ex_{source}},$$

where $\sum Ex_{product}$ and $\sum Ex_{source}$ have to be specified for each type of system.

5.6 EXERGY TRANSFER BY HEAT

Figure 5.4 shows the schematic diagram of exergy transfer by heat.

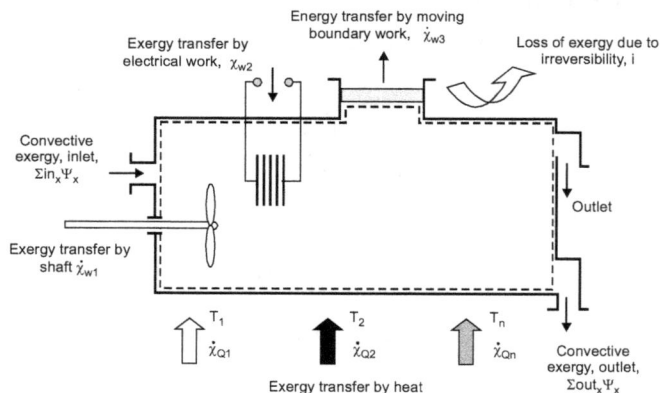

FIGURE 5.4 Exergy transfer by heat.

REVIEW QUESTIONS

1. Discuss various types of thermodynamic properties and their correlations.
2. Explain the following:
   a. Virial equation of state.
   b. Van der Waals equation of state.

3. Derive Maxwell's relation in terms of:

   a. Energy and entropy

   b. Enthalpy

   c. Helmholtz free energy

   d. Gibbs free energy

   e. Ground potential

4. Explain the following with the schematic diagrams:

   a. Exergy transfer by heat.

   b. Flow exergy of matter.

5. Define exergy efficiency.

NUMERICAL EXERCISES

1. Determine the reversible work and the irreversibility associated with a process in which a metallic ball of 2 kg is cooled down from 300°C to 150°C by keeping it in the open space at 20°C. Specific heat of the metal may be taken as 0.45 kJ/kg·K. **(143 kJ, 55 kJ)**

2. 5 kg of air at 550 K and 4 bar is enclosed in a closed vessel.

   (*i*) Determine the availability of the system if the surrounding's pressure and temperature are 1 bar and 290 K, respectively.

   (*ii*) If the air is cooled at a constant pressure to the atmospheric temperature, determine the availability and effectiveness.

   (576.7 kJ, 64.8%)

3. 5 Kg of ice at −10°C is kept in atmosphere which is at 30°C. Calculate the change in entropy of universe when it melts and comes into thermal equilibrium with the atmosphere. Take latent heat of fusion as 335 kJ/kg and specific heat of ice as half that of water. **(0.652 kJ)**

4. 0.25 kg/s of water is heated from 30°C to 60°C by hot gases that enter at 180°C and leave at 80°C. Calculate the mass flow rate of gases when its C_p = 1.08 kJ/kg·K. Find the enthalpy change of water and of hot gases. Take the specific heat of water as 4.186 kJ/kg·K. Also find the increase of unavailable energy if the ambient temperature is 27°C.

   (0.0988 kJ, 23.436 kJ)

THERMODYNAMIC CYCLES

6.1 INTRODUCTION

The majority of power-producing machines work on thermodynamic cycles and therefore research and development work in the area of thermodynamics has assumed to gain very high importance.

Every power engine operates as a steady-state device and does not store power. The working fluid passes through a series of processes forming a closed-loop cycle. The cycle of processes is repeated again and again as the engine operates.

Thermodynamic cycles are classified as gas cycles and vapor cycles. In gas cycles, the fluid remains in gaseous phase, whereas the face may change in vapor cycles. If steam is the working fluid, it will be gas in some processes and liquid in other processes.

Thermodynamic cycles are also classified as an open cycle and a closed cycle. In a closed cycle, the same working fluid is moved from initial to fluid state and then recirculates. In the case of open cycles, the working fluid is changed in each cycle and not being recirculated. In a car engine, the burnt gases can be exhausted and a fresh mixture of air and fuel is changed at the end of every cycle.

There are engines where the working fluid does not undergo an entire thermodynamic cycle.

Thermodynamic cycles are also categorized as per their utility.

1. Power cycles

2. Heat pump and refrigeration cycles.

1.   **Power cycles:** thermodynamic cycles that are used in engine for power production are called power cycles. Power cycles may be either gas power cycles or vapor power cycles.

The gas power cycles are further classified as follows:

a. Stirling cycle

b. Dual cycle

c. Carnot gas power cycle

d. Diesel cycle

e. Ericsson cycle

f. Otto cycle

g. Brayton cycle

The vapor power cycles are further classified as follows:

a. Carnot vapor power cycle

b. Rankine cycle

c. Reheat cycle

d. Regenerative cycle

Carnot, reheat, and Rankine cycles will be discussed in detail.

2.   **Heat pump and refrigeration cycles:** these thermodynamic cycles are used for heat pump and refrigeration. They may be either vapor cycles or air cycles depending upon the working fluid used.

6.2   CYCLE DEFINITION

A thermodynamic cycle consists of a series of processes and normally the working fluid returns to its original state. It is also called as a closed cycle.

Figure 6.1 shows a thermodynamic cycle that is considered to be a closed system. A thermodynamic cycle may not be a closed cycle. For instance, a jet engine repeatedly receives fresh ambient air and

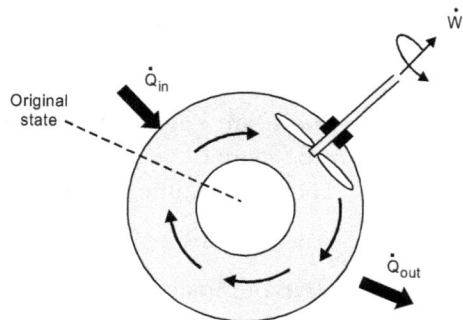

FIGURE 6.1 Thermodynamic cycle.

releases hot burnt gases to the atmosphere. The burnt gas is cooled down in the atmosphere to ambient temperature.

The concept of thermodynamic cycle, as shown in Figure 6.2, can be explained as follows:

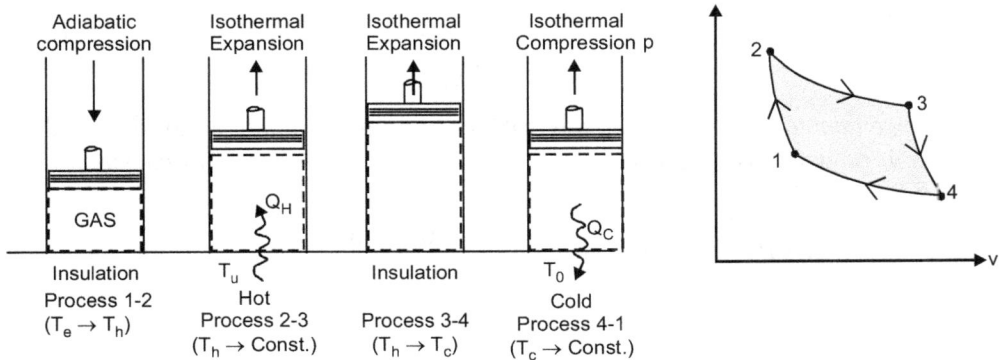

FIGURE 6.2 The system's cycle with the addition of heat.

Consider a fluid contained in a piston–cylinder apparatus that undergoes heat interaction with two heat reservoirs at different temperatures. The fluid exercises the following four processes and regains its original state.

1–2: adiabatic compression to raise the temperature to high-temperature reservoir.

2–3: the fluid at the temperature of high-temperature reservoir is isothermally heated.

3–4: adiabatic expansion to the temperature of lower-temperature reservoir.

4–1: isothermal cooling of the gas to the initial pressure and the temperature.

The cycle produced by these four processes is called the Carnot cycle.

6.3 PERFORMANCE PARAMETERS

A few performance parameters of power cycles are explained below.

Thermal effectiveness: it is the limit of the quantity of heat entering the engine which is converted to the work of the engine.

$$Thermal\ efficiency = \frac{Network\ of\ cycle}{Heat\ added\ in\ cycle} \tag{6.1}$$

Ratio of back work: the ratio of the pump work ($-W$) required to the turbine work ($+W$) produced.

$$Back\ work\ ratio = \frac{W_{pump}}{W_{turbine}} \qquad (6.2)$$

Usually, ratio of back work is small indicating the pump work ($-W$) required to the turbine work ($+W$) produced.

Work ratio:

$$Work\ ratio = \frac{W_{net}}{W_{turbine}} \qquad (6.3)$$

Specific steam consumption (SSC): it is the amount of steam required to produce unit power. It is usually expressed as kg/kWh and has a numerical value from 3 to 5 kg/kWh.

$$Specific\ steam\ consumption = \frac{3600}{W_{net}}, \frac{kg}{kWh} \qquad (6.4)$$

6.4   CARNOT VAPOR POWER CYCLE

The Carnot cycle consists of four reversible processes: isothermal addition of heat, isentropic expansion, isothermal discharge of heat, and isentropic compressor as shown in Figure 6.3.

Carnot cycle is a perfect cycle having maximum thermodynamic effectiveness. Carnot vapor cycle contains vapor as the working fluid. The *T-S* and *P-V* diagrams of a Carnot cycle are plotted in Figure 6.4. The Carnot cycle works as a closed system.

A cylinder–piston arrangement or a turbine with a gas or a vapor as the working fluid is used for Carnot cycle.

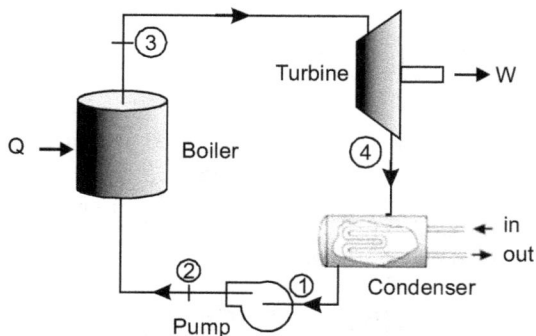

FIGURE 6.3 Steady-flow Carnot engine.

It consists of (i) two constant pressure processes (2–3) and (4–1) and (ii) two adiabatic processes without friction (1–2) and (3–4). The four steps can be described in detail as follows:

1. *Process (2–3)*: At temperature T_2 and from dryness practices X_2, 1 kg of hot water is heated to saturated steam pressure P_3 and temperature T_3. The pressure ($P_2 - P_3$) and the temperature ($T_2 - T_3$) remain constant during this process.

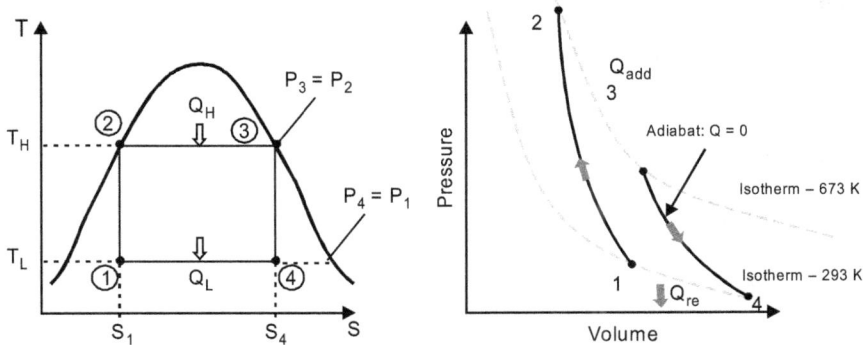

FIGURE 6.4 Arrangement of Carnot cycle.

2. *Process (3–4):* The vapor is expanded by isentropic process to temperature T_4 and pressure P_4. Point "4" indicates the state of vapor after expansion.

3. *Process (4–1):* Heat is rejected at a constant temperature $(T_4 = T_1)$ and pressure $(P_4 = P_1)$. The vapor is condensed into liquid.

4. *Process (1–2):* The wet vapor at "1" is compressed isentropically untill the vapor reaches its initial condition of P_2 and T_2. Then the cycle gets completed.

Heat added: $Q_H = T_H(s_3 - s_2)$
Heat rejected: $Q_L = T_L(s_4 - s_1)$

6.4.1 Plant Components

1. Pump

Water enters the pump at state 1 and is pumped isentropically to the working pressure of the boiler. The water temperature rises through this isentropic compression process due to a small reduction of specific volume of water.

2. Boiler

Water enters the boiler as a condensed liquid at state 2 and leaves as a super-heated/saturated vapor at state 3. Heat is transferred to the water at a constant pressure. This heat may be generated in a nuclear reactor, burning gases, or any other sources of heat. When the vapor reaches the superheated state, the boiler is called a steam generator.

3. Turbine

Superheated steam enters the turbine, and expands isotropically producing work by rotating a shaft coupled to an electric generator.

4. Condenser

Condenser is a heat exchanger where vapor is condensed by abstracting the heat at a constant pressure P_4.

Vapor leaves the condenser as a saturated liquid and enters the pump to complete the cycle.

6.4.2 Thermal Analysis of Carnot Cycle

$$\text{Thermal efficiency} = \frac{\text{Net work}}{\text{Heat added}} \tag{6.5}$$

$$\text{Net work} = \text{turbine work} - \text{pumping work}$$

Net work per unit mass flow:

$$W_{net} = W_T - W_P$$
$$W_{net} = (h_3 - h_4) - (h_2 - h_1) \tag{6.6}$$

Heat added in boiler:

$$Q_{add} = (h_3 - h_2) \tag{6.7}$$

Heat rejected in condenser:

$$Q_{rejected} = (h_4 - h_1) \tag{6.8}$$

Thermal efficiency:

$$\therefore \qquad \eta_{carnot} = \frac{W_{net}}{Q_{add}} \tag{6.9}$$

$$\therefore \qquad \eta_{carnot} = \frac{(h_3 - h_4) - (h_2 - h_1)}{(h_3 - h_2)} \tag{6.10}$$

$$\therefore \qquad \eta_{carnot} = 1 - \frac{Q_{rejected}}{Q_{add}} \tag{6.11}$$

The heat addition and rejection can be expressed in terms of temperature and entropy as follows:

Since:

$$Q = T\Delta s$$

$$\therefore \qquad Q_{add} = T_2(s_3 - s_2) \tag{6.12}$$

$$\therefore \qquad Q_{rejected} = T_1(s_4 - s_1) \tag{6.13}$$

Also:

$$s_1 = s_2 \quad \text{and} \quad s_3 = s_4$$

$$\eta_{carnot} = 1 - \frac{T_L}{T_H} \tag{6.14}$$

or
$$\eta_{carnot} = 1 - \frac{T_1}{T_2}$$

6.4.3 Limitations of Carnot's Cycle

Thermodynamically, the Carnot cycle is simple, and when the values of T_2 and T_1 are known, the cycle reaches maximum thermal effectiveness. But it is very difficult to use this cycle in practice due to the following limitations:

1. The process of condensation of vapor to saturation condition is difficult.

2. To find the state 4 in the beginning of condensation is difficult.

3. The critical temperature of water vapor is fixed at 374°C, therefore, maximum possible temperature is bounded.

4. It is difficult to work with the superheated steam because providing the superheat at a constant temperature rather than constant pressure is difficult.

6.5 PRINCIPLES OF THE CARNOT CYCLE

The Carnot cycle is the best reversible cycle.

But there is irreversibility in the real process. Therefore, actual cycles are less efficient than the Carnot cycle.

The Carnot cycle has four processes as shown in Figure 6.5 for a closed system of piston–cylinder arrangement.

1–2: Reversible isothermal expansion process

The gas expands at a constant temperature T_H and the heat Q_H is transferred to gas.

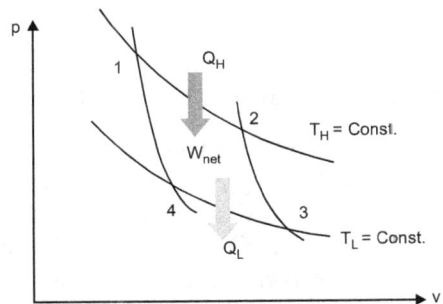

FIGURE 6.5 *v-p* graph of Carnot cycle.

2–3: *Reverse adiabatic expansion process*

The gas expands by the reversible process, and the temperature of gas reduces to T_L from T_H.

3–4: *Reverse isothermal compression*

Heat is rejected to the surroundings from the system (reversible) at a constant gas temperature T_2.

4–1: *Reversible adiabatic compression process*

The temperature of gas increases to T_H from T_L.

Carnot cycle works between two separate temperatures T_H and T_2.

For all reversible and irreversible heat cycles,

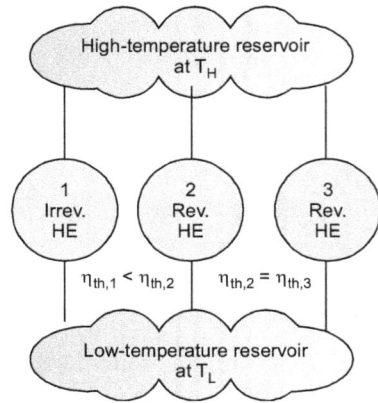

FIGURE 6.6 Carnot principle.

$$\eta_{th} = 1 - \frac{Q_L}{Q_H} \tag{6.15}$$

For the Carnot cycle,

$$\eta_{th,Carnot} = 1 - \frac{T_L}{T_H} \tag{6.16}$$

The efficiency of the Carnot cycle is always more than the efficiency of an irreversible (real) working between the same two reservoirs as shown in Figure 6.6.

$$\eta_{th} = \begin{cases} < \eta_{th,rev} & \text{irreversible cycle} \\ = \eta_{th,rev} & \text{reversible cycle} \\ > \eta_{th,rev} & \text{impossible cycle} \end{cases} \tag{6.17}$$

6.5.1 The Reverse Carnot Cycle

The reverse Carnot cycle is used for refrigeration. Q_H represents the amount of heat rejected to a reservoir at high temperature, whereas Q_L represents the amount of heat received from a reservoir at low temperature. Work entering the system is important to complete the cycle.

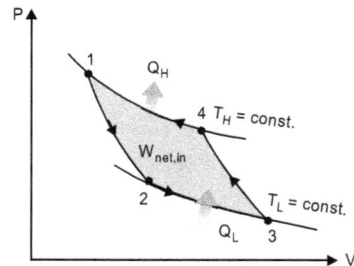

FIGURE 6.7 P-V diagram of the reversed Carnot cycle.

The *V-P* diagram of the reversed Carnot cycle is similar to the usual Carnot cycle, but the directions of the process are reversed as shown in Figure 6.7.

6.5.2 Carnot Cycle for Heat Pump and Refrigerator

A heat pump or refrigerator working on a reversed Carnot cycle is called as refrigerator or a Carnot heat pump.

The coefficient of performance (COP) of any refrigerator or heat pump is calculated as follows:

$$COP_R = \frac{1}{\dfrac{Q_H}{Q_L} - 1}$$

and

$$COP_{HP} = \frac{1}{1 - \dfrac{Q_L}{Q_H}}$$

The COP of the reversible refrigerators or heat pumps is calculated as follows:

$$COP_{R,rev} = \frac{1}{\dfrac{T_H}{T_L} - 1}$$

and

$$COP_{HP,rev} = \frac{1}{1 - \dfrac{T_L}{T_H}}$$

Similar to the heat engine, one can conclude:

$$COP_R = \begin{cases} < COP_{R,rev} & \text{irreversible refrigerator} \\ = COP_{R,rev} & \text{reversible refrigerator} \\ > COP_{R,rev} & \text{impossible refrigerator} \end{cases} \qquad (6.18)$$

6.5.3 Carnot Cycle for Heat Engine

All heat engines working on the principles of Carnot cycle are called the Carnot heat engines. The thermal effectiveness of any heat engine (irreversible or reversible) is calculated as follows:

$$\eta_{th} = 1 - \frac{Q_L}{Q_H}, \qquad (6.19)$$

where Q_H is the heat transferred to the engine from the reservoir at a high temperature T_H, and Q_L is the heat rejected to the reservoir at low temperature T_L.

For the reversible heat engines, the ratio of heat transfer with reservoir is equal to the ratio of the temperatures of the two reservoirs.

Therefore, efficiency of a Carnot engine, or some reversible heat engine, will be as follows:

$$\eta_{th,rev} = 1 - \frac{T_L}{T_H} \tag{6.20}$$

This is the maximum efficiency of heat engine working between thermal reservoirs T_H and T_L as shown in Figure 6.8.

All irreversible (real) heat engines working in the range of two temperature limits (T_H and T_L) have lesser efficiency.

The actual thermal efficiencies of the heat engines working between the same ranges of temperatures (Figure 6.9) are as follows:

$$\eta_{th} = \begin{cases} < \eta_{th,rev} & \text{irreversible engine} \\ = \eta_{th,rev} & \text{reversible engine} \\ > \eta_{th,rev} & \text{impossible engine} \end{cases} \tag{6.21}$$

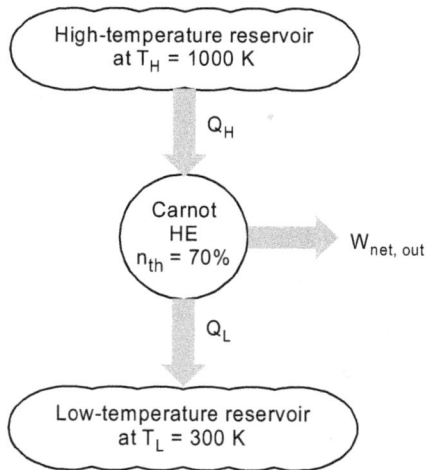

FIGURE 6.8 Carnot heat engine.

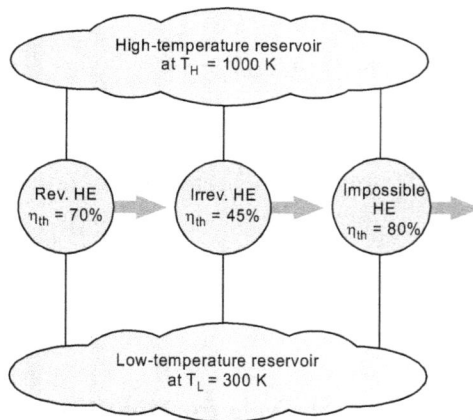

FIGURE 6.9 Carnot principles.

EXAMPLE 6.1

A Carnot cycle operates in vapor in the range of pressure from 7 kPa to 7 MPa. Calculate the work of turbine, thermal efficiency, and work of compressing of vapor.

Solution:

Enthalpy at state 3, $h_3 = h_g$ at 7 MPa

$$h_3 = 2772.1 \text{ kJ/kg}$$

Entropy at state 3, $s_3 = s_g$ at 7 MPa

$$s_3 = 5.8133 \text{ kJ/kg.K}$$

Enthalpy at state 2,

$$h_2 = h_f \text{ at 7 MPa}$$

$$h_2 = 1267 \text{ kJ/kg}$$

Entropy at state 2,

$$s_2 = s_f \text{ at 7 MPa}$$

$$s_2 = 3.1211 \text{ kJ/kg.K}$$

For process 3–4,

$s_4 = s_3$, consider fraction of aridness at state 4 to be x_4

$$s_4 = s_3 = s_f + (s_{fg} * x_4) \text{ at 7 kPa}$$

$$5.813300 = 0.556400 + x_4.7.723700$$

$$x_4 = 0.680600$$

For enthalpy at state 4:

$$h_4 = h_f + (h_{fg} * x_4) \text{ at 7 kPa}$$

$$= 162.6000 + (0.680600 \times 2409.5400)$$

$$h_4 = 1802.5300 \text{ kJ/kg}$$

Fraction of dryness at point 1 is x_1

For process 1–2,

$$s_2 = s_1 = s_f + (s_{fg} * x_1) \text{ at 7 kPa}$$

$$3.121100 = 0.556400 + (x_1.7.723700)$$

$$x_1 = 0.332100$$

For enthalpy at state 1:

$$h_1 = h_f + (h_{fg} * x_1) \text{ at 7 kPa}$$

$$= 162.6000 + (0.332100 \times 2409.5400)$$

$$h_4 = 962.81 \text{ kilojoule/kilogram}$$

$$\text{Thermal Efficiency} = \frac{\text{Network}}{\text{Heat added}}$$

Work of expansion (work of turbine) $= -(h_4 - h_3)$

$$= -(1802.5300 - 2772.100)$$

$$= 969.5700 \text{ kJ/kg}$$

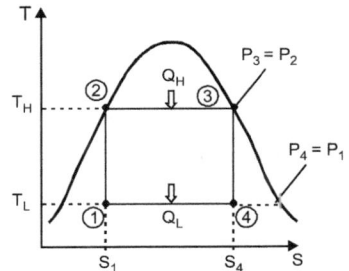

T-s diagram with states labeled ②, ③, ①, ④, showing Q_H, Q_L, T_H, T_L, $P_3 = P_2$, $P_4 = P_1$, s_1, s_4, S.

FIGURE 6.10

$$\text{Work of Compressing} = -(h_1 - h_2)$$
$$= -(962.8100 - 126700)$$
$$= 304.1900 \text{ kJ/kg (positive sign)}$$

$$\text{Amount of heat added to a system} = -(h_2 - h_1)$$
$$= -(1267.00 - 2772.100)$$
$$= 1505.1 \text{ kJ.kg (positive sign)}$$

$$\text{Overall work} = -(h_4 - h_3) + (h_1 - h_2)$$
$$= 969.5700 - 304.1900 = 665.3800 \text{ kJ/kg}$$

$$\text{Thermal Efficiency} = \frac{665.38}{1505.1} = 0.4421 \text{ or } 44.21\% \text{ Ans.}$$

6.6 IDEAL RANKINE CYCLE

Some of the limitations of the Carnot cycle can be removed by superheating the vapor in the boiler and a complete condensing in the condenser. As shown in Figure 6.11, the cycle is called the Rankine cycle. This is the ideal cycle for steam power plants.

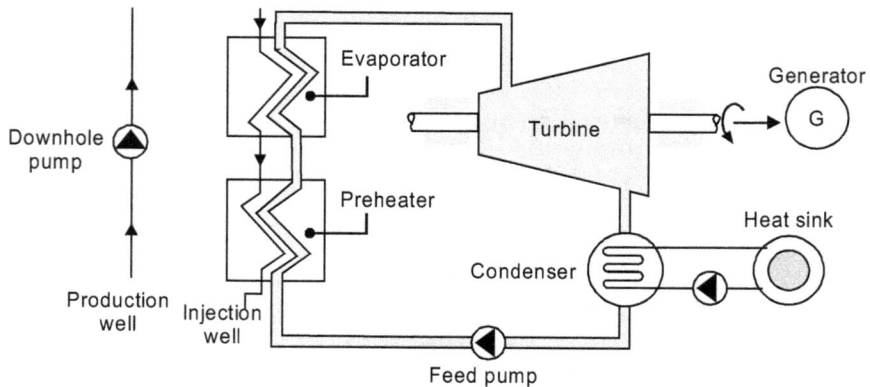

FIGURE 6.11 Schematic layout of ideal Rankine cycle.

The T-S diagram of ideal Rankine cycle is shown in Figure 6.12.

The ideal Rankine cycle consists of the following four processes:

Process 1–2: isentropic compressor in a pump.
Process 2–3: addition of heat in a boiler at a constant pressure.

Process 3–4: development of isentropic work in an engine.

Process 4–1: rejection of heat in a condenser at a constant pressure.

6.6.1 Analysis of the Ideal Rankine Cycle

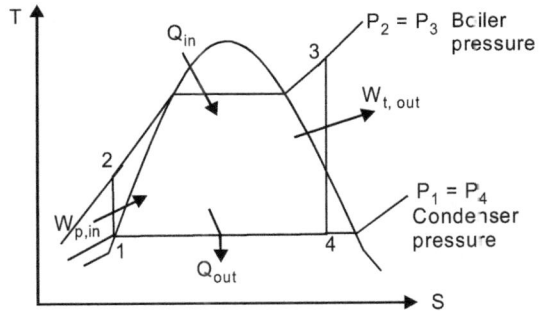

FIGURE 6.12 *T-S diagram of an ideal Rankine cycle.*

The four machines of Rankine cycle (condenser, turbine, boiler, and pump) have constant flow. Rankine cycle is considered as a constant-flow process.

There is no work for the condenser and the boiler. The pump and the turbine are taken as isentropic.

1–2: Pump work:

$$W_{\text{pump}} = W_p = (h_2 - h_1) \text{ kJ/kg}, \tag{6.22}$$
$$= v_f (p_2 - p_1) \text{ kJ/kg}$$

where p_2 and p_1 are in kPa.
The values of v_f and p_2 can be obtained from this steam tables.

$$\text{Pump power} = \dot{m} w_p \text{ in kW} \tag{6.23}$$

2–3: Heat supplied in boiler:

$$q_{\text{supply}} = (h_3 - h_2) \text{ kJ/kg} \tag{6.24}$$

or, $$Q_{\text{supply}} = \dot{m}(h_3 - h_2) \text{ KW} \tag{6.25}$$

3–4: Turbine work:

$$w_T = (h_3 - h_2) \text{ kJ/kg} \tag{6.26}$$

$$\text{Turbine power } W_T = \dot{m}(h_3 - h_2) \text{ kW}, \tag{6.27}$$

where p_1 = pressure of condenser or low pressure of cycle, h_1 = enthalpy from steam table at p_1, p_2 = pressure of boiler or pressure at turbine inlet or high pressure of cycle, h_2 = enthalpy from steam table at P_2, $\dot{m}$ = the mass flow rate of steam in kg/sec.

It is possible to use Mollier diagram to get h_2 and h_1.

4–1: Condensation process at constant pressure:

$$q_{rejected} = (h_4 - h_1) \text{ KJ/kg} \tag{6.28}$$

$$Q_{rejected} = \dot{m}(h_4 - h_1) \text{ kW}$$

$$h_1 = h_f \text{ at low pressure } p_1 \tag{6.29}$$

(i) Net work:

$$W_{net} = W_T - W_P \tag{6.30}$$

(ii) Thermal efficiency:

Ratio of the network to the heat supplied.

$$
\eta_{thermal} \text{ (or) } \eta_{rankine} \text{ (or) } \eta_{cycle} = \frac{w_{net}}{q_{supply}}
$$

$$
= \frac{w_T - w_P}{q_{supply}} \tag{6.31}
$$

$$
= \frac{(h_3 - h_4) - (h_2 - h_1)}{(h_3 - h_2)}
$$

(iii) Specific steam consumption:

$$
SSC = \frac{3600}{W_{net}} \quad \frac{\text{kg}}{\text{kW-hr}} \tag{6.32}
$$

$$
\text{Work ratio} = \frac{W_{net}}{W_T} \tag{6.33}
$$

$$
\text{The back work ratio: } bwr = \frac{W_P}{W_T} = \frac{h_2 - h_1}{h_3 - h_4} \tag{6.34}
$$

EXAMPLE 6.2

An ideal Rankine cycle has steam as the working fluid. Saturated steam enters the turbine at a pressure of 8000 kPa, whereas saturated liquid leaves the condenser at a pressure of 8 kPa. The cycle power output is 100,000 KW.
Calculate:

a. **The thermal efficiency of the cycle.**

b. **The back work ratio of the cycle.**

c. **The flow rate of steam, in kg/h.**

d. **The flow rate of condenser cooling water, in kg/h, if cooling water enters the condenser at 15°C and leaves at 35°C.**

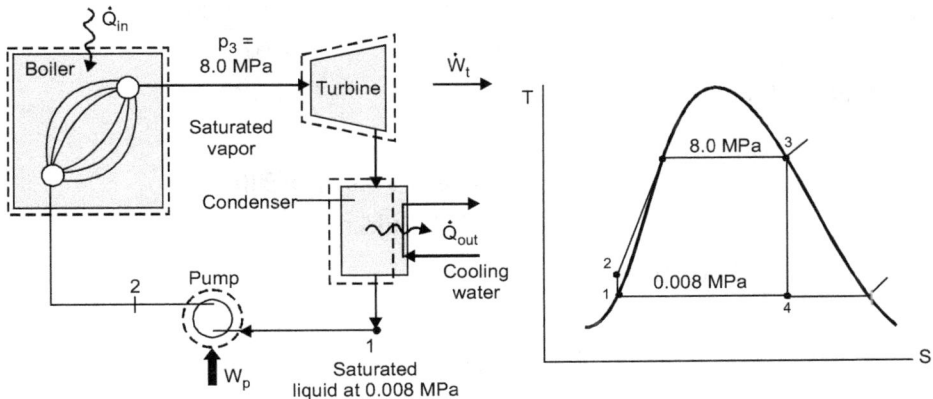

FIGURE 6.13 Steam power plant.

Solution:

State 1 (pump):

$P_1 = 0.008$ MPa,

$h_1 = h_{f,p_1} = 173.88$ kJ/kg

Saturated liquid

$v_1 = v_{f,p_1} = 1.0084 \times 10^{-3} \text{m}^3 / \text{kg}$

State 2 (boiler):

$p_2 = 8.0$ MPa

$s_2 = s_1$

$w_{pump} = v_1(P_2 - P_1)$

$= 1.0084 \times 10^{-3}(8 - 0.008)$

$= 8.06$ kJ/kg

$h_2 = h_1 + W_{pump}$

$= 173.88 + 8.06 = 181.94$ kJ/kg

State 3 (turbine):
Saturated vapor

$p_3 = 8.0$ MPa $\quad h_3 = 2758.0$ kJ/kg

$s_3 = 5.7432$ kJ/kg.K

$$p_4 = 0.008 \text{ MPa}$$

$$s_3 = s_4 \left. \begin{array}{l} p_4 = 0.008 \text{ MPa} \\ s_3 = s_4 \end{array} \right\}_{s_4 = s_3} = 5.7432 \text{ kJ/kg.K}$$

$$x_4 = \frac{s_4 - s_f}{s_{fg}} = \frac{5.7432 - 0.5926}{7.6361}$$

State 4 (condenser):
$$x_4 = 0.6756$$
$$h_4 = h_f + x_4 h_{fg}$$
$$= 173.88 + 0.6745 \times 2403.1$$
$$h_4 = 1794.8 \text{ kJ/kg}$$

a. Thermal efficiency:

$$\eta_{\text{thermal}} = \frac{W_{\text{net}}}{Q_{\text{supply}}}$$

$$= \frac{-(h_4 - h_3) + (h_1 - h_2)}{-(h_2 - h_3)}$$

$$\eta_{\text{thermal}} = \frac{-(h_4 - h_3) + (h_1 - h_2)}{-(h_2 - h_3)}$$

$$\eta_{\text{thermal}} = \frac{-(1794.8 - 2758.0) + (173.88 - 181.94)}{-(181.94 - 2758.8)}$$

$$\eta_{\text{thermal}} = 0.371 = 37.1\%$$

b. Back work ratio:

$$bwr = \frac{W_P}{W_T} = \frac{h_2 - h_1}{h_3 - h_4}$$

$$bwr = \frac{W_P}{W_T} = \frac{181.94 - 173.88}{2758.0 - 1794.8}$$

$$bwr = 8.37 \times 10^{-3} = 0.84\% \text{ **Ans.**}$$

c. Mass flow rate of steam:

$$\dot{m} = \frac{W_{\text{cycle}}}{(h_3 - h_4) - (h_2 - h_1)}$$

$$\dot{m} = \frac{(100) \times 10^3 \times 3600}{(963.2) - (8.06)}$$

$$\dot{m} = 3.77 \times 10^5 \text{ kg/h} \quad \text{**Ans.**}$$

d. Flow rate of condenser cooling water:

$$\dot{m} = (h_4 - h_1)$$
$$= \dot{m}_w C_w (T_{wout} - T_{win})$$
$$3.77 \times 10^5 (179.4.8 - 173.88) = \dot{m}_w \times 4.2(35 - 15)$$
$$\dot{m}_w = 7.2 \times 10^5 \text{ kg/h} \quad \textbf{Ans.}$$

6.6.2 Comparison between Carnot and Rankine Cycles

i. For the same temperature limits, Rankine cycle gives higher output than the Carnot cycle. Rankine cycle needs a lesser flow rate of vapor to obtain a known output. Rankine cycle needs higher rates of heat transfer in boiler and condenser.

ii. Efficiency is less than that of Carnot cycle. The efficiency of Rankine cycle can be more than that of Carnot cycle, if superheated steam is used.

iii. The compressor power is very high in Carnot cycle as compared to pumping power in Rankine cycle.

Figure 6.14 gives comparison of SSC and efficiency of two cycles.

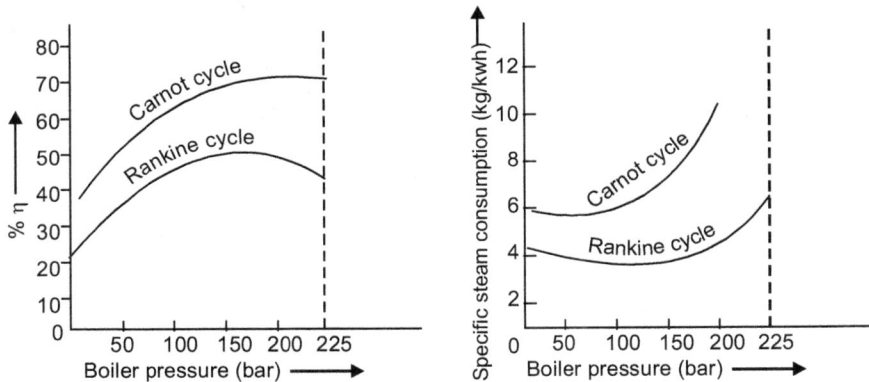

FIGURE 6.14 Specific steam consumption and thermal efficiency of Carnot and ideal Rankine cycles.

6.6.3 Effect of Operating Conditions on Rankine Cycle Efficiency

The Rankine cycle efficiency can be increased by the following factors:

1. Lowering the condenser pressure (lowers $T_{low, avg}$)

The effect of lower pressure of condenser on the Rankine cycle efficiency is shown in *T-S* diagram (Figure 6.15). The shaded area in Figure 6.15 shows the increase of power output by lowering the condenser pressure. The heat input shown by the area under the curve 2'-2 increases but very little. The thermal efficiency of the cycle increases with the decrease in condenser pressure.

The vapor power cycles work in a closed loop and the condenser pressure is lower than the atmospheric pressure to increase the thermal efficiency of the cycle.

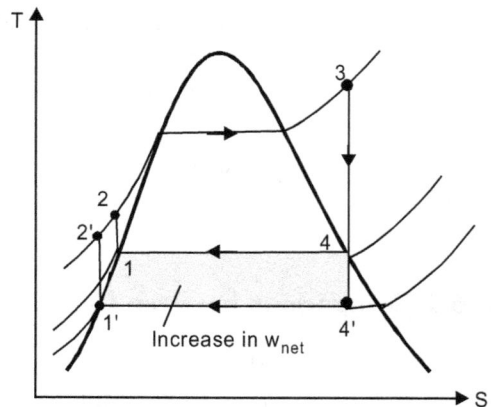

FIGURE 6.15 Result of lowering the condenser pressure on the ideal Rankine cycle.

2. Superheating the steam to high temperature (increases $T_{high,\ avg}$)

The steam temperature can be increased by superheating without the increase in boiler pressure as indicated by the shaded zone in Figure 6.16. The increase in the heat input and the higher work output are shown by the area under the curve 3-3'. This leads to higher thermal efficiency.

There is an additional advantage of superheating the vapor to higher temperatures as the dryness fraction of steam at the turbine's exit improves.

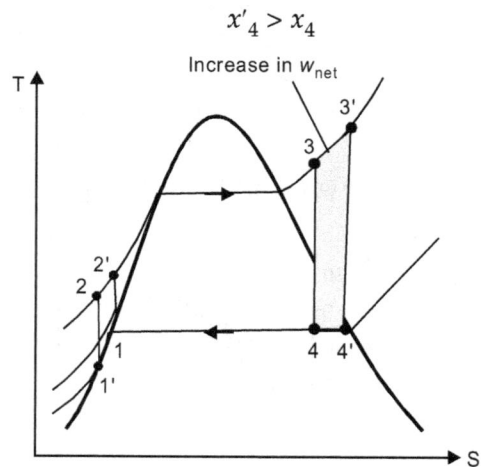

FIGURE 6.16 Ideal Rankine cycle showing the influence of superheating the vapor to higher temperatures.

3. Raising the boiler pressure (increases $T_{high,\ avg}$)

The temperature of heat-addition process in the boiler can be increased by raising the working pressure of the boiler. But the dryness fraction of steam at the turbine's exit decreases as point 4 moves to the left.

These methods can be used to increase the thermal efficiency of the Rankine cycle.

    i.   Use of dual vapor.

    ii.   Extraction of water from steam.

iii. Feed water reheating or regeneration.

iv. Reheating of steam.

6.6.4 Regeneration Cycle

In Rankine cycle, the condensate at low temperature takes an irreversible addition of heat in a hot boiler thereby cycle efficiency decreases. The condensate can be heated with the steam extracted from the turbine before sending to the boiler.

This heating is called regeneration heating and the cycle is called regeneration cycle.

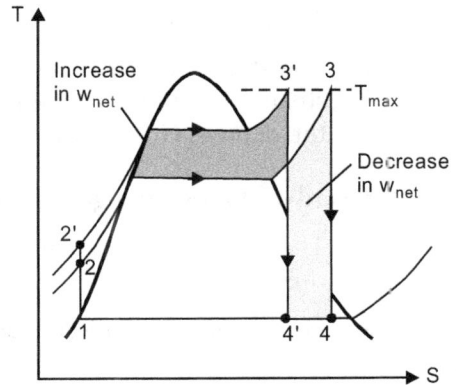

FIGURE 6.17 Effect of increasing the boiler pressure on the ideal Rankine cycle.

The regenerative heating is always utilized and a number of the heaters are provided. The process of steam extraction is called bleeding of steam. The number of feed water heaters (FWHs) depends on the size of plant.

Eight to nine heaters may be used for large thermal power plants.

A FWH is a heat exchanger to heat condensate by steam. There are two types of FWHs:

1. Open or direct contact type.

2. Closed type with:

    a. Drains cascaded backward

    b. Drains pumped forward.

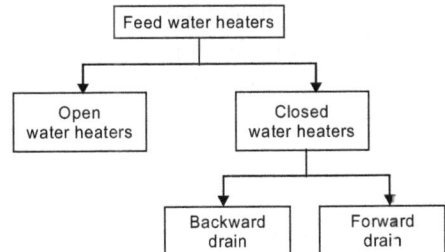

FIGURE 6.18 Types of feed water heaters.

Advantages of regeneration cycle:

1. The heat supply in the boiler is decreased.

2. The temperature difference in the boiler is reduced.

3. The boiler heating approaches to reversible process.

4. The temperature of heat addition in boiler increases.

5. The thermal efficiency depends on the temperature of heat supply and therefore increases.

6. The size of condenser is reduced.

7. The steam flow in turbine becomes wet and water has to be separated by steam bleeding, and the need for water separation decreases along with the decrease in the corrosion of turbine blades.

Disadvantages of regeneration cycle:

1. The plant becomes complicated increasing capital cost and decreasing reliability.

2. Heaters need additional maintenance.

3. Boiler size increases.

4. The heaters are expensive and the gain in thermal efficiency is not much as compared to additional cost.

6.6.4.1 Open FWHs

The advantages of two types of FWHs are given in the following table:

Open FWHs	Closed FWHs
1. Simple and inexpensive.	1. Complex and expensive because of the internal tubing network.
2. More efficient heat transfer due to direct contact.	2. Less efficient since the two streams are not allowed to come in direct contact.
3. One pump can be used for many FWHs	3. One pump is used for each FWH

In an open-type FWH, the feed water from condenser and bled steam from turbine are directly mixed. The schematic diagram of a power plant along with T-S graph is shown in Figure 6.19.

In a cycle of Rankine ideal regeneration, vapor enters the turbine at the pressure of the boiler and decreases isentropically to a moderate pressure. Some steam is removed at this condition and same to the heater. The rest of vapor stays to expand isentropically to the pressure of condenser. This vapor goes to the condenser as a saturated liquid at the pressure of the condenser. The feed water enters the pump at an isentropic condition, wherever it is condensed to the pressure of heater of feed water (stage 2) and is in flying to the provide for water warmer, wherever feed water combines with the vapor extracted from the turbine.

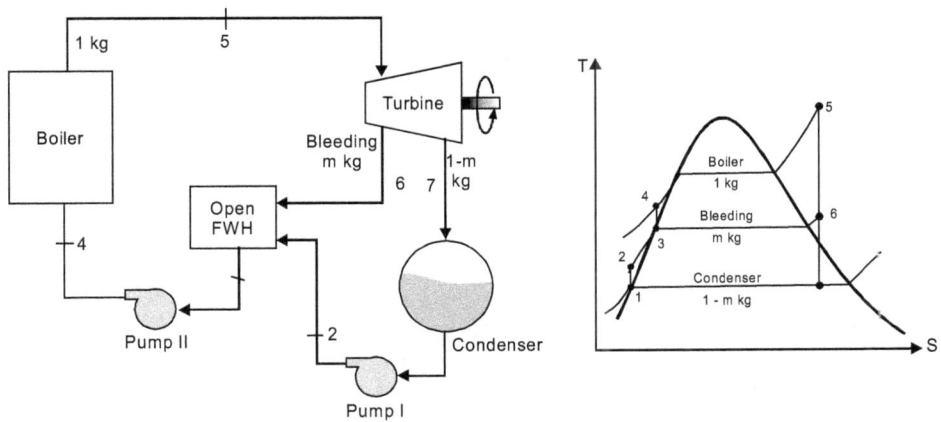

FIGURE 6.19 Ideal regeneration Rankine cycle with open feed water heater.

Thermal analysis of power plant with open FWH:

$$q_{in} = h_5 = h_4$$
$$q_{out} = (1-m)(h_7 - h_1)$$
$$W_{turb.out} = (h_5 - h_6) + (1-m)(h_6 - h_7)$$
$$W_{pump.in} = (1-m)W_{pumpI,in} + W_{pumpII,in}$$

Where:

$$W_{pumpI.in} = V_1(P_2 - P_1)$$
$$W_{pumpII.in} = V_3(P_4 - P_3)$$

Energy balance of FWH:

For 1 kg of working fluid,

Inlet of energy = outlet of energy

$$m = \frac{\text{mass of steam bled}}{\text{mass of steam circulated}} \qquad (6.35)$$

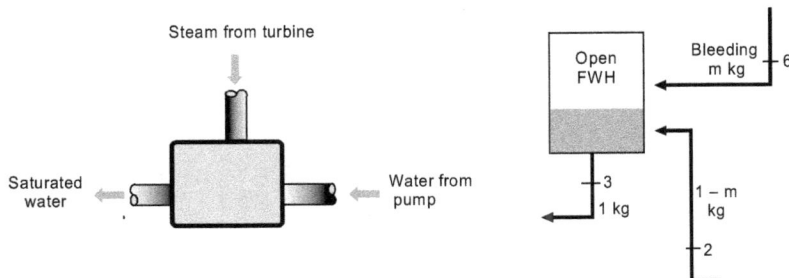

FIGURE 6.20

Energy entering the heater = Energy leaving the heater

$$m = \frac{h_3 - h_2}{h_6 - h_2} = \text{Mass of steam bled}$$

$$m \cdot h_6 + (1 - m)h_2 = 1 \times h_3$$

$$m \cdot h_6 + h_2 - mh_2 = h_3$$

$$m(h_6 - h_2) - h_3 - h_2 \tag{6.36}$$

EXAMPLE 6.3
In an ideal regeneration Rankine cycle of steam power plant working with one open feed water heaters, steam enters the turbine at 600.00°C and 15000 kPa and condensed at 10.000 kPa. Steam is heated at a pressure of 1.2 MPa and enters the feed water heater. Calculate the fraction of steam bled and the thermal efficiency of the cycle.

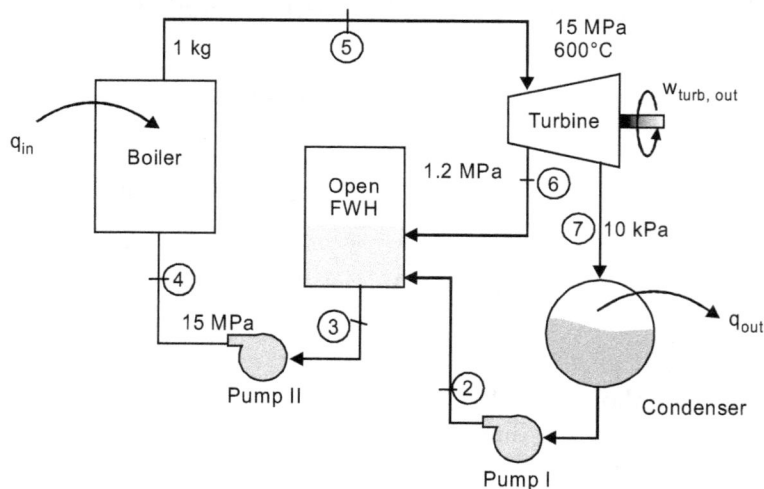

FIGURE 6.21 Schematic diagram of Example 6.3.

Solution:

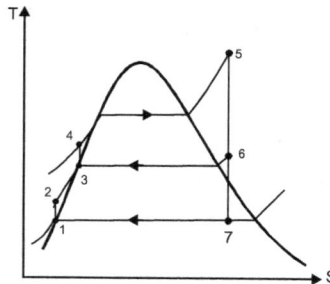

FIGURE 6.22 T-S diagram.

Condenser outlet:

$$\left. \begin{array}{l} P_1 = 10 \text{ kPa} \\ \text{Sat. Liquid} \end{array} \right\} h_1 = h_{f@10 \text{ kPa}} = 191.81 \text{ kJ/kg}$$

$$v_1 = v_{f@10 \text{ kPa}} = 0.00101 \text{ m}^3/ \text{kg}$$

Pump outlet:

$$P_2 = 1.2 \text{ MPa}$$

$$s_2 = s_1$$

$$W_{\text{pumpI.in}} = v_1 (P_2 - P_1)$$

$$= (0.00101 \text{ m}^3/ \text{kg})[(1200 - 10) \text{ kPa}] \left(\frac{1 \text{ kJ}}{1 \text{ kPa} \cdot \text{m}^3} \right)$$

$$= 1.20 \text{ kJ/kg}$$

$$h_2 = h_1 + W_{\text{pumpI.in}} = (191.81 + 1.20) \text{ kJ/kg}$$

$$= 193.01 \text{ kJ/kg}$$

FWH outlet:

$$\left. \begin{array}{l} P_3 = 1.2 \text{ MPa} \\ \text{Sat. Liquid} \end{array} \right\} v_3 = v_{f@1.2 \text{ MPa}} = 0.001138 \text{ m}^3/\text{kg}$$

$$h_3 = h_{f@1.2 \text{ MPa}} = 798.33 \text{ kJ/kg}$$

Boiler inlet:

$$P_4 = 15 \text{ MPa}$$

$$s_4 = s_3$$

$$W_{\text{pumpII.in}} = v_3 (P_4 - P_3)$$

$$= (0.001138 \text{ m}^3/\text{kg})[15,000 - 1200] \text{ kPa} \left(\frac{1 \text{ kJ}}{1 \text{ kPa} \cdot \text{m}^3} \right)$$

$$= 15.70 \text{ kJ/kg.}$$

$$h_4 + h_3 + W_{\text{pumpII.in}} = (798.33 + 15.70) \text{ kJ/kg} = 814.03 \text{ kJ/kg}$$

Turbine inlet:

$$\left. \begin{array}{l} P_5 = 15 MPA \\ T_5 = 600°X \end{array} \right\} h_5 = 3583.1 \text{ kJ/kg}$$

$$s_5 = 6.6796 \text{ kJ/kg.K}$$

Bled steam:

$$P_6 = 1.2 \text{ MPa} \atop s_6 = s_5 \Big\} h_5 = 2860.2 \text{ kJ/kg}$$

$$(T_6 = 218.4°C)$$

Condenser inlet:

$$P_7 = 10 \text{ kPa}$$

$$s_7 = s_5 \quad x_7 = \frac{s_7 - s_f}{s_{fg}} = \frac{6.6796 - 0.6492}{7.4996} = 0.8041$$

$$h_7 = h_f + x_7 h_{fg} = 191.81 + 0.8041(2392.1) = 2115.3 \text{ kj/kg}$$

$$m\, h_6 + (1-m)h_2 = 1(h_3)$$

$$m = \frac{h_3 - h_1}{h_6 - h_2} = \frac{798.33 - 193.01}{2860.2 - 193.01} = 0.2270 \text{kg.} \quad \textbf{Ans.}$$

$$\therefore q_{in} = h_5 - h_4 (3583.1 - 814.03) \text{ kJ/kg} = 2769.1 \text{ kJ/kg}$$

$$q_{out} = (1-m)(h_7 - h_1)1 \text{ kg}$$

$$= (1 - 0.2270)(2115.3 - 191.81) \text{ kj/kg} = 1486.9 \text{ kJ/kg.}$$

$$\therefore \eta_{th} = 1 - \frac{q_{out}}{q_{in}} = 1 - \frac{1486.9 \text{ kJ/kg}}{2769.1 \text{ kJ/kg}} = 0.463 \text{ or } 46.3\% \quad \textbf{Ans.}$$

6.6.4.2 Closed FWHs

Closed FWHs are shell-and-tube heat exchangers in which feed water temperature increases as the bled steam condenses on the outer of the tubes with feed water flows inside the tubes.

The two steams are at different pressures and do not mix.

The bled steam condenses in the closed feed water while heating the feed water from the pump.

The heated feed water is sent to the boiler and the condensates from the FWH are also pumped to the boiler.

There are two types of closed feed water heating schemes as shown in Figure 6.23.

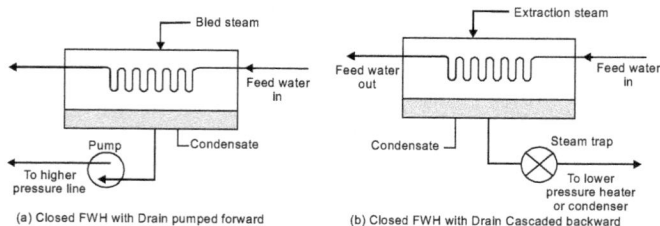

(a) Closed FWH with Drain pumped forward   (b) Closed FWH with Drain Cascaded backward

FIGURE 6.23 Feed water heating schemes.

i.   Closed FWH with drains cascaded backward

- This scheme of FWHs is simple and mostly used.
- It is shell and tube heat exchanger.
- The feed water flows in the tubes and the bled steam flows in the shell.
- Only one pump is needed as the steam does not mix with the feed water.
- If decreator is used, another pump may be used before the boiler feed pump.
- The bled steam is drained back to previous low-pressure FWH by reducing the pressure or led back to the condenser.
- Terminal Temperature Difference (TTD) is defined as the difference between temperature of bled steam entering the feed water and the sub-cooled water temperature. TTD is shown in Figure 6.24, as:

TTD = Saturation temperature of bled steam − exit water temperature. The TTD is positive and is often of the order of 0°C–5°C.

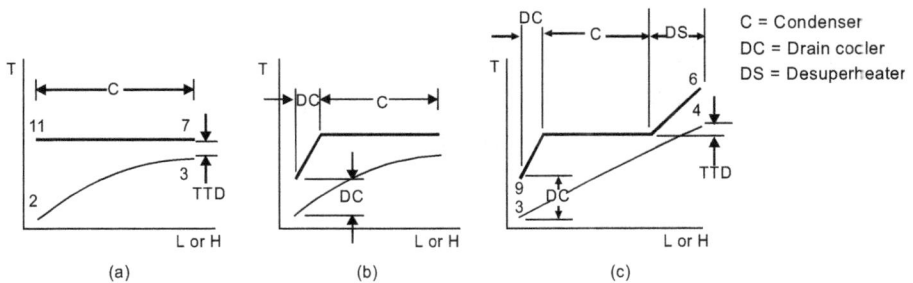

FIGURE 6.24 Terminal temperature difference.

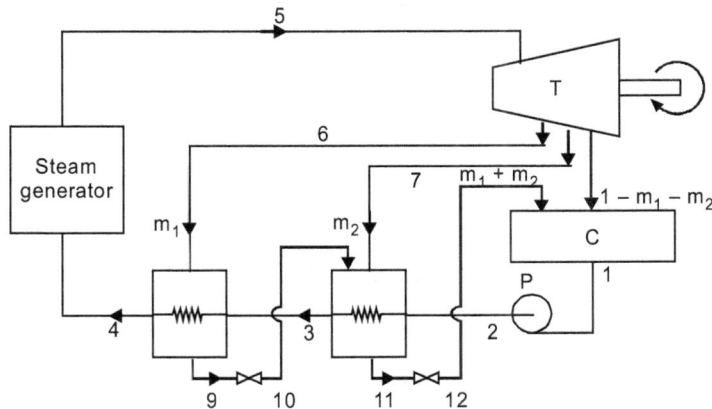

FIGURE 6.24(A) Power plant with closed FWH and drains cascaded backward.

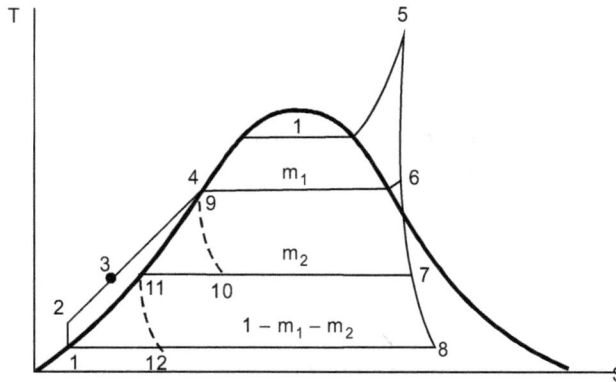

FIGURE 6.25 Closed FWH with drains cascaded backward.

m_1 = mass flow rate of bled steam for the high pressure closed FWH

m_2 = mass flow rate of bled steam for the low pressure closed FWH

$m_s = m_6 + m_7 + m_8$ = total steam mass flow

Energy balance of the high-pressure FWH:

The energy balance of the high-pressure FWH is given as follows:

Energy entering the heater = Energy leaving the heater

$$m_1 h_6 + h_3 = m_1 h_9 + h_4 \quad m_1(h_6 - h_9) = h_4 - h_3$$

For throttling process, enthalpy is constant.

$$\therefore \quad h_9 = h_{10}$$

Energy balance of the low-pressure FWH:

Energy entering the heater = Energy leaving the heater

$$m_1 h_{10} + m_2 h_7 + h_2 = (m_1 + m_2)h_{11} + h_3$$

$$m_1(h_{10} - h_{11}) + m_2(h_7 - h_{11}) = h_3 - h_2$$

For throttling process, enthalpy is constant.

$$h_{11} = h_{12}$$

Thermal analysis of cycle:

For boiler: $\qquad q_{in} = (h_5 - h_4)$ \hfill (6.37)

For condenser:

$$q_{out} = (1 - m_1 - m_2)(h_8 - h_1) + (m_1 + m_2)(h_{12} - h_1) \hfill (6.38)$$

For pump: $\qquad W_{in} = (h_2 - h_1) = v_1(P_2 - P_1)$ \hfill (6.39)

For turbine:
$$W_{out} = (h_5 - h_6) + (1 - m_1)(h_6 - h_7) + (1 - m_1 - m_2)(h_7 - h_g) \qquad (6.40)$$

For cycle:
$$W_{net} = W_{out} - W_{in} \qquad (6.41)$$

Thermal efficiency:
$$\eta = \frac{W_{net}}{q_{in}} \qquad (6.42)$$

EXAMPLE 6.4

In a steam power plant working on the ideal regenerative Rankine cycle with two closed feed water heaters, steam enters at 600.00°C and pressure of 15000 kPa and it is condensed at 10.00 kPa. At 4000 kPa, a small amount of steam is bled from the turbine. At 500 kPa, steam enters closed feed water heater. Calculate the cycle's thermal efficiency and rate of mass flow of steam entering the two closed FWHs.

Solution:

Pump inlet: $\quad P_1 = 10 \text{ kPa } h_1 = h_{f@P_1} = 191.81 \text{ kJ/kg}$

Saturated liquid: $\quad v_1 = v_{f,P_1} = 0.00101 \text{ m}^3/\text{kg}$

$$s_1 = 0.6492 \text{ kJ/kg K}$$

Pump outlet: $\quad p_2 = p_5 = 15 \text{ MPa}$

$$s_2 = s_1$$

$$\therefore \quad W_{pump} = v_1(p_2 - p_1) = 0.00101(15000 - 10)$$

$$= 15.14 \text{ kJ/kg}$$

$$h_2 = h_1 + W_{pump}$$

$$= 191.81 + 15.14 = 206.95 \text{ kJ/kg}$$

Turbine inlet:

$$p_5 = 15 \text{ MPa } \quad h_5 = 3582.3 \text{ kJ/kg}$$

$$T_s = 600°C \quad s_5 = 6.6775 \text{ kJ/kg.K}$$

$$s_5 = s_6 = s_7 = s_8$$

Bled steam:

$$p_6 = 4 \text{ MPa } \quad h_6 = 3152 \text{ kJ/kg}$$

$$s_6 = 6.6775 \text{ kJ/kg.K at } T_6 = 375°C$$

FWH inlet:

$$p_7 = 0.5 \text{ MPa } h_f = h_{11} = 640.21 \text{ kJ/kg}$$

$$s_7 = 6.6775 \text{ kJ/kg.K} \quad h_{fg} = 2108.47 \text{ kJ/kg}^{-1}$$

$$\text{Mixture } s_f = 1.860600 \text{ kJ.kg}^{-1}.\text{K}^{-1}$$

$$s_{g \cdot f} = 4.9606 \text{ kJ.kg}^{-1}.\text{K}$$

$$x_7 = \frac{s_7 - s_f}{s_{fg}} = \frac{6.6775 - 1.8606}{4.9606}$$

$$h_7 = h_f + x_7 h_{fg}$$

$$= 640.21 + 0.971 * 2108.47$$

$$h_7 = 2687.5 \text{ kJ/kg}$$

Condenser inlet:

$$p_8 = 10 \text{ kPa } h_f = 191.81 \text{ kJ/kg}$$

$$s_8 = s_7 = 6.6775 \text{ kJ/kg.K } h_{fg}$$

$$= 2392.82 \text{ kJ.kg}^{-1}$$

$$\text{Mixture } s_f = 0.649200 \text{ kJ.kg}^{-1}.\text{K}^{-1}$$

$$s_{g \cdot f} = 7.501 \text{ kJ.kg}^{-1}.\text{k}^{-1}$$

$$x_8 = \frac{s_8 - s_f}{s_{fg}} = \frac{6.6775 - 0.4692}{7.501}$$

$$x_8 = 0.803$$

$$h_8 = h_f + x_8 h_{fg}$$

$$= 191.81 + 0.803 \times 2392.82$$

$$h_8 = 2113.2 \text{ kJ/kg}$$

$$p_9 = 4 \text{ MPa } h_9$$

$$= 1087.29 \text{ kJ/kg}$$

$$\text{Sat.} T_9 = 250.4°\text{C}$$

Assume TTD = 2°C

$$\text{TTD} = T_{11} - T_3 (T_{11} = T_{\text{sat}}, 500 \text{ kPa.})$$

$$2 = 151.86 - T_3$$

$$T_3 = 149.86°\text{C} \quad T_3 \approx 150°\text{C}$$

$$T_3 = 150°\text{C} \quad h_3 = h_f = 632.18 \text{ kJ/kg}$$

At
$$T_3 = 150°C \ \ h_3 = h_f = 632.18 \text{ kJ/kg}$$

$$\text{Assuem TTD} = 2°C$$

$$\text{TTD} = T_9 - T_4 \ (T_9 = T_{\text{sat}}, 4 \text{ MPa.})$$

$$2 = 250.4 - T_4$$

$$T_4 = 248.4°C \ \ T_4 \approx 248°C$$

At
$$T_4 = 248°C \ \ h_4 = h_f = 1074 \text{ kJ/kg}$$

The energy balance of the high-pressure FWH is given as follows:

Energy entering the regenerator = Energy leaving the regenerator

$$m_1 h_6 + h_3 = m_1 h_9 + h_4 m_1 (h_6 - h_9) = h_4 - h_3$$

$$m_1 (3152 - 1087.29) = 1074 - 632.18$$

$$m_1 = 0.214 \text{ kg}$$

The energy balance of the low-pressure FWH is given as follows:

Energy entering the regenerator = Energy leaving the regenerator

$$m_1 h_{10} + m_2 h_7 + h_2 = (m_1 + m_2) h_{11} + h_3$$

$$(h_{10} - h_{11}) + m_2 (h_7 - h_{11}) = h_3 - h_2$$

For throttling process, enthalpy is constant.

$$h_9 = h_{10}$$

$$h_{11} = h_{12}$$

$$h_{11} = h_{\text{sat}}, \ 500 \text{ kPa}$$

$$= 640.21 \text{ kJ/kg}$$

$$0.214(1087.29 - 640.21) + m_2(2687.5 - 640.21)$$

$$= 638.18 - 206.95$$

$$m_2 = 0.161 \text{ kg} \quad \textbf{Ans.}$$

$$q_{in} = (h_5 - h_4) q_{in}$$

$$= (3582.3 - 1074)$$

$$q_{in} = 2503.3 \text{ kJ/kg}$$

$$q_{out} = (1 - m_1 - m_2)(h_8 - h_1) + (m_1 + m_2)(h_2 - h_1)$$

$$q_{out} = (1 - 0.214 - 0161)(2113.2 - 191.81)$$

$$+ (0.214 + 0.161)(640.21 - 191.81)$$

$$q_{out} = 1369 \text{ kJ/kg}$$

$$\eta = \frac{W_{net}}{q_{in}} = 1 - \frac{q_{out}}{q_{in}}$$

$$\eta = 1 - \frac{1369}{2508.3}$$

$$\eta = 0.454$$

$$= 45.4\% \quad \textbf{Ans.}$$

i. Closed FWH with drains pumped forward

The field water is heated by bled steam without mixing. The pressures of two streams are different. The condensate of bled steam leaves the heater as saturated water. Figure 6.26 shows an ideal regenerative Rankine cycle with a closed FWH with drains pumped forward.

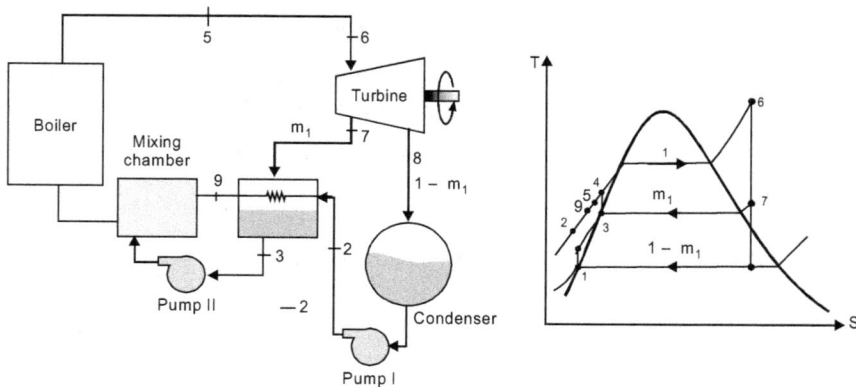

FIGURE 6.26 The ideal regenerative Rankine cycle with closed FWH with drains pumped forward.

Energy and mass balance of FWH

m_1 = mass flow rate of bled steam for the closed FWH

$m_s = m_7 + m_8$

The energy balance of the closed FWH is given as follows:

E inlet FWH = E outlet FWH

$$m_1 h_7 - (m_1 - 1) h_2 = m_1 h_3 - (m_1 - 1) h_9 \tag{6.43}$$

Thermal analysis of cycle:
For boiler:

$$q_{in} = (h_6 - h_5) \tag{6.44}$$

For condenser:

$$q_{out} = (1 - m_1)(h_8 - h_1) \tag{6.45}$$

For pump:

$$w_{in} = (1 - m_1)(h_2 - h_1) \qquad (6.46)$$

Net work:

$$W_{net} = W_{out} - W_{in} \qquad (6.47)$$

Thermal efficiency:

$$\eta = \frac{W_{net}}{q_{in}} \qquad (6.48)$$

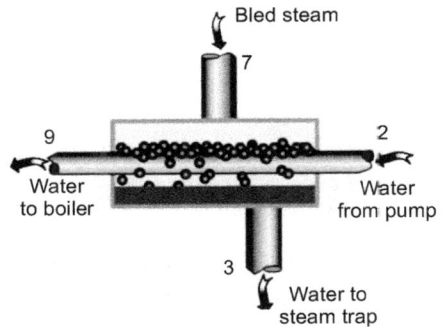

FIGURE 6.27 Closed FWH.

EXAMPLE 6.5

Repeat the solution of Example 6.4 using the same data with FWHs installed in forward pumped drain arrangement.

Solution: From previous example:

$$h_1 = h_f, p_1 = 191.81 \text{ kJ/kg}$$
$$h_2 = 206.95 \text{ kJ/kg}$$
$$h_7 = 3582.3 \text{ kJ/kg}$$
$$h_8 = 2113.2 \text{ kJ/kg}$$
$$h_9 = 3152 \text{ kJ/kg}$$
$$h_{10} = 2687.5 \text{ kJ/kg}$$
$$h_{11} = 1087.29 \text{ kJ/kg}$$
$$h_{13} = 640.21 \text{ kJ/kg}$$

Assume

$$\text{TTD} = 2°\text{C to fined } h_3 \text{ and } h_5$$
$$\text{TTD} = T_{13} - T_3 (T_{11} = T_{\text{sat,500 kPa.}})$$
$$2 = 151.86 - T_3$$
$$T_3 = 149.86°\text{C} \quad T_3 \approx 150°\text{C}$$

At

$$T_3 = 150°\text{C} \quad h_3 = h_1 = 632.18 \text{ kJ/kg}$$
$$\text{TTD} = T_{11} - T_5 (T_{11} = T_{\text{sat}}, 4_{\text{MPa.}})$$
$$2 = 250.4 - T_4$$
$$T_5 = 248.4°\text{C} \quad T_5 \approx 248°\text{C}$$

At

$$T_5 = 248°\text{C} \quad h_5 = h_f = 1074 \text{ kJ/kg}$$

For low-pressure FWH, the pump work:

$$W_{pump} = v_{11}(p_7 - p_{11}) = 0.001252(15000 - 4000)$$
$$= 13.77 \text{ kJ/kg}$$
$$h_{12} = h1_1 + W_{pump}$$
$$= 1087.29 + 13.77 = 1101.06 \text{ kJ/kg}$$

The energy equilibrium of the high-pressure FWH is given as follows:
E incoming to the regenerator = E exiting from the regenerator

$$m_1 h_9 - (m_1 - 1)h_4 = m_1 h_{11} - (m_1 - 1)h_5$$

The energy balance of the low-pressure FWH is given as follows:
E incoming to the regenerator = E exiting from the regenerator

$$m_1 h_9 - (m_1 - 1)h_4 = m_1 h_{11} - (m_1 - 1)h_5$$

The energy balance of the low-pressure FWH is given as follows:
Energy entering the regenerator = Energy parting the regenerator

$$m_2 h_{10} - (1 - m_1 - m_2)h_2 = (1 - m_1 m_2)h_3 + h_2 h_{13}$$
$$m_1 = 0.1776 \text{ kg}$$
$$m_2 = 0.1414 \text{ kg}$$
$$h_4 = 636.284 \text{ kJ/kg}$$
$$h_6 = 1078.806 \text{ kJ/kg}$$
$$q_{in} = (h_7 - h_6) \quad q_{in} = (3582.3 - 1078.806)$$
$$q_{in} = 2503.3 \text{ kJ/kg}$$
$$q_{out} = (1 - m_1 - m_2)(h_8 - h_1)$$
$$q_{out} = (1 - 0.1776 - 0.1414)(2113.2 - 191.81)$$
$$q_{out} = 1308.46 \text{ kJ/kg}$$
$$\eta = \frac{W_{net}}{q_{in}}$$
$$= 1 - \frac{q_{out}}{q_{in}}$$
$$= 1 - \frac{1308.46}{2503.5}$$
$$\eta = 0.477 = 47.7\%. \textbf{ Ans.}$$

6.6.5 The Placement of FWHs

For maximum thermal efficiency, it is very important to decide the pressure of bled steam from turbine.

$$\Delta T_{opt} = \frac{T_B - T_C}{n+1}$$

6.6.6 Reheat Cycle

The thermal effectiveness of the cycle can be increased by the following methods:

i. Regenerative feed water heating.
ii. Binary vapor cycle.
iii. Reheating of steam.
iv. Removal of water.

Reheating system

Figure 6.28(b) shows schematic diagram and T-S diagram of an ideal reheat Rankine cycle.

Steam is expanded in the high-pressure turbine isentropically to an inter-moderate pressure and sent back to the boiler for reheating to initial temperature. Steam is expanded isentropically in the low-pressure turbine to the condenser pressure.

$$\therefore q_{in} = q_{primary} + q_{reheat} = (h_3 - h_2) + (h_5 - h_4)$$
$$w_{turb,\,out} = w_{turb,\,I} + q_{turb,\,II} = (h_3 - h_4) + (h_5 - h_6) \tag{6.49}$$

The improvement in thermal efficiency by reheating depends on the reheat pressure with reference to the inlet stem pressure.

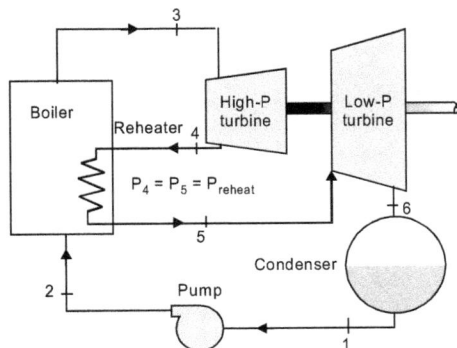

FIGURE 6.28(A) Ideal reheat Rankine cycle.

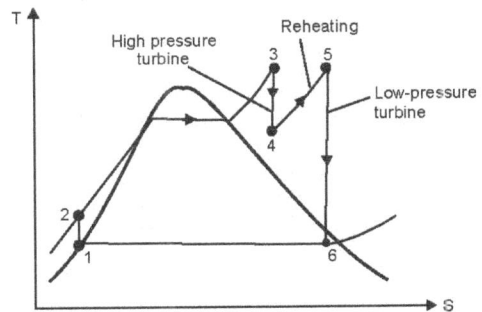

FIGURE 6.28(B) T-S diagram of ideal reheat Rankine cycle.

Figure 6.29 shows the reheat pressure versus cycle efficiency.

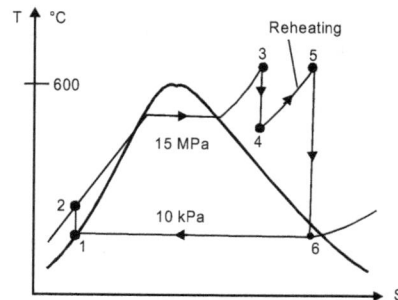

Advantages of reheating:

1. Output power increases.

2. Heat supply increases.

3. Thermal efficiency increases.

4. The wetness drops and corrosion of blade becomes less. This leads to the increase in turbine life.

Disadvantages of reheating:

1. Added equipment and piping increase plant cost and operational reliability decreases.

2. The increase in thermal efficiency may not offset the disadvantages of plant complexity.

Pressure of Condenser: 12.700 mm Hg
Inlet and reheat temperature: 427.00°C

FIGURE 6.29 Reheat pressure versus cycle efficiency.

EXAMPLE 6.6

In an ideal reheat Rankine cycle steam power plant, steam vapor enters the turbine at a temperature of 600.00°C and pressure of 15000 kPa, and is condensed at 10.000 kPa. The vapor moisture at the exit of turbine should not exceed 10.400%, calculate the following:

a. Reheat pressure.

b. Thermal efficiency of cycle.

Assume the steam is reheated to the initial temperature.

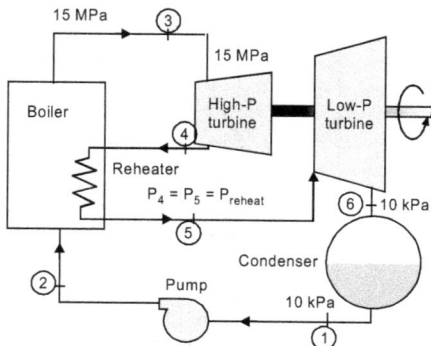

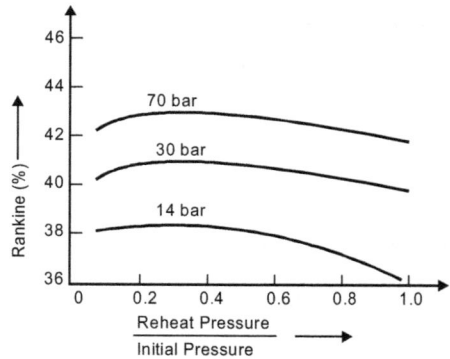

FIGURE 6.30 Reheat Rankine cycle.

Solution:

a. The reheat pressure is determined from the requirement that the entropies at states 5 and 6 are the same:

Turbine outlet:

$$P_6 = 10 \text{ kPa}$$

$$x_6 = 0.896 \text{ (sat.miture)}$$

$$S_6 = s_f + x_6 s_{fg}$$

$$= 0.6492 + 0.896(7.4996)$$

$$= 7.3688 \text{ kJ/kg.K}$$

But

$$h_6 = h_f + x_6 h_{fg}$$

$$\therefore \quad h_6 = 191.81 + 0.896(2392.1)$$

$$= 2335.1 \text{ kJ/kg}$$

Thus,

Turbine inlet:

$$\left. \begin{matrix} T_5 = 600°C \\ s_5 = s_6 \end{matrix} \right| \quad \begin{matrix} P_5 = 4.0 \text{ MPa} \\ h_5 = 3674.9 \text{ kJ/kg} \end{matrix}$$

Therefore, steam should be reheated at a pressure of 4 MPa or lower to prevent a moisture content above 10.4%.

b. To determine the thermal efficiency, the enthalpies at all other states must be known:

Condenser outlet:

$$P_1 = 10 \text{ kPa} \left| h_1 = h_{f @ 10 \text{ kPa}} = 191.81 \text{ kJ/kg} \right.$$

$$\text{Sat. liquid} \left| v_1 = v_{f ® 19 \text{ kPa}} = 0.00101 \text{ m}^3\text{/kg} \right.$$

Boiler inlet:

$$P_2 = 15 \text{ MPa}$$

$$s_2 = s_1$$

$$W_{\text{pump.in}} = v_1(p_2 - p_1) = (0.00101 \text{ m}^3\text{/kg}) \times [(15.000 - 10 \text{ kPA})] \left(\frac{1 \text{ kJ}}{1 \text{ kPa} \cdot \text{m}^3} \right)$$

$$= 15.14 \text{ kJ/kg.}$$

$$h_2 = h_1 + w_{\text{pump.in}}$$

$$= (191.81 + 15.14) \text{ kJ/kg} = 206.95 \text{ kJ/kg}$$

High-pressure turbine inlet:

$$P_3 = 15 \text{ MPa} \left.\right\} h_3 = 3583.1 \text{ kJ/kg}$$
$$T_3 = 600°\text{C} \left.\right\} s_3 = 6.6796 \text{ kJ/kg-K}$$

High-pressure turbine outlet:

$$P_4 = 4 \text{ MPa} \left.\right\} h_4 = 3155.0 \text{ kJ/kg}$$
$$s_4 = s_3 \left.\right\} (T_4 = 375.5°\text{C})$$

Thus

$$q_{in} = (h_3 - h_2) + (h_5 - h_4)$$
$$= (3583.1 - 206.95) \text{ kJ/kg} + (3674.9 - 3155.0) \text{ kJ/kg}$$
$$= 3896.1 \text{ kJ/kg}$$
$$q_{out} = h_6 - h_1 = (2335.1 - 191.81) \text{ kJ/kg} = 2143.3 \text{ kJ/kg}$$

and

$$\eta = 1 - \frac{q_{out}}{q_{in}} = 1 - \frac{2143.3 \text{ kJ/kg}}{3896.1 \text{ kJ/kg}} = 0.450 \text{ or } 45.0\% \quad \textbf{Ans.}$$

6.6.7 Cogeneration Cycles

Cogeneration is the simultaneous generation of electricity and steam (or heat) in a single power plant. Chemical industries, paper mills, and municipalities that use district heating need to process heat or steam as well as electricity. These industries and municipalities can produce electricity more conveniently and cheaply by cogeneration.

The cogeneration plan efficiency is calculated as follows:

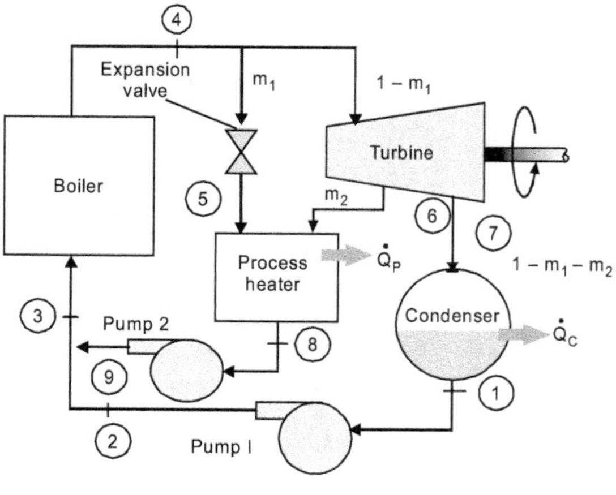

FIGURE 6.31 Cogeneration plant with adjustable loads.

$$\eta_{co} = \frac{\text{electric energy generated} + \text{process heat}}{\text{heat added to the plant}}$$

A cogeneration plant with adjustable load is shown in Figure 6.31.

The boiler steam is used for both electric power generators and heat. Factor of utilization $\in_u$ of cogeneration plant is calculated as follows:

$$\in_u = \frac{\text{Net work output} + \text{Process heat delivered}}{\text{Total heat input}}$$

$$= \frac{W_{\text{net}} + Q_p}{\dot{Q}_{\text{in}}} \ldots \tag{6.50}$$

Let m_4 = total mass flow rate of steam in kg/s

$$= m_5 + m_6 + m_7$$

$$q_{\text{in}} = (h_4 - h_3)$$

$$q_{\text{out}} = (1 - m_1 - m_2)(h_7 - h_1) \tag{6.51}$$

$$q_p = m_1 h_5 + m_2 h_6 - (m_1 + m_2)h_8 \tag{6.52}$$

$$W_{\text{in}} = (1 - m_1 - m_2)(h_2 - h_1) + (m_1 + m_2)(h_9 - h_8) \tag{6.53}$$

$$W_{\text{out}} = (1 - m_1)(h_4 - h_6) + (1 - m_1 + m_2)(h_9 - h_7) \tag{6.54}$$

$$W_{\text{net}} = W_{\text{out}} - W_{\text{in}} \tag{6.55}$$

Or

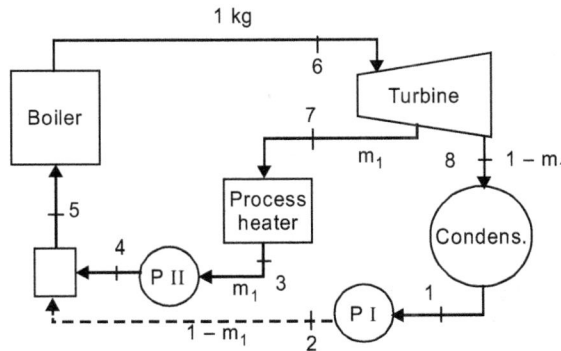

FIGURE 6.32

$$q_{\text{in}} = (h_6 - h_5) \tag{6.56}$$

$$q_{\text{out}} = (1 - m_1)(h_8 - h_1) \tag{6.57}$$

$$q_p = m_1 h_7 - m_1 h_3 \tag{6.58}$$

$$W_{in} = (1-m_1)(h_2-h_1)+(m_1)(h_4-h_3) \quad\quad\quad (6.59)$$

$$W_{out} = (h_6-h_7)+(1-m_1)(h_7-h_8) \quad\quad\quad (6.60)$$

$$W_{net} = W_{out} - W_{in} \quad\quad\quad (6.61)$$

EXAMPLE 6.7

In a cogeneration plant, boiler produces steam at temperature of 450.00°C and pressure of 10000 kPa at a rate of 5.00 kg·s^{-1}. Steam expands to 500 kPa in the turbine, then it is supplied to the process heater. Before entering the boiler, it is heated in a feed water heater. The condenser works at 20.00 kPa.

    a. Calculate the rate of power generation.

    b. If only 60% of steam is used as process heat and the condenser receives the remaining steam, calculate the rate of power generation.

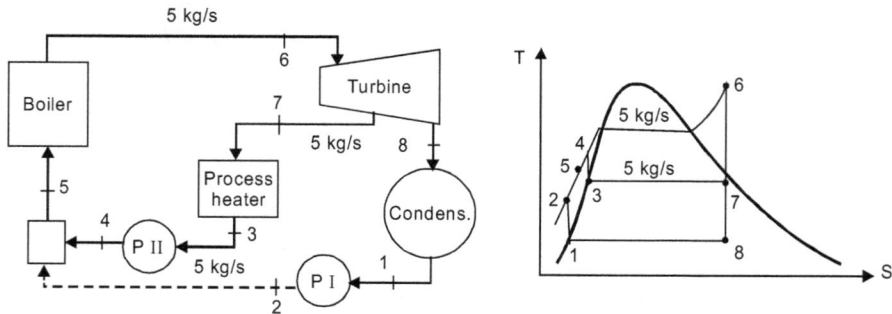

FIGURE 6.33 Schematic diagram of Example 6.7.

Solution: (a)

For pump P-I inlet:

$$P_1 = 20 \text{ kPa } h_1 = h_{f,p_1} = 251.42 \text{ kJ/kg}$$

$$v_1 = v_{f,P_1} = 0.001017 \text{m}^3\text{/kg}$$

Pump outlet:

$$W_{p_1} = v_1(P_2-P_1) = 0.001017(10000-20)$$

$$= 10.15 \text{ kJ/kg}$$

$$h_2 = h_1 + W_{p_1}$$

$$= 251.42 + 10.15 = 261.57 \text{ kJ/kg}$$

For pump *P-H* inlet:

$$P_3 = 0.5 \text{ MPa} \quad h_3 = h_{f,p_3} \; 640.09 \text{ kJ/kg}$$

$$v_3 = v_{f,p_3} = 0.001093 \text{ m}^3/\text{kg}$$

Pump outlet:

$$W_{p_1} = v_3(P_4 - P_3)$$

$$= 0.001093(10000 - 500)$$

$$= 10.38 \text{ kJ/kg}$$

$$h_4 = h_3 + W_{p_{11}}$$

$$= 640.09 + 10.38 = 650.47 \text{ kJ/kg}$$

Turbine inlet:

$$P_6 = 10 \text{ MPa} \quad h_6 = 3242.4 \text{ kJ/kg}$$

$$T6 = 450°C \quad s_6 = 6.4219 \text{ kJ/kg.K}$$

Bled steam:

$$p_7 = 0.5 \text{ MPa}, \; h_f = 640.09 \text{ kJ/kg}$$

$$s_7 = s_6 = 6.4219 \text{ kJ/kg.K}$$

$$h_{fg} = 2108.0 \text{ kJ/kg}$$

$$\text{Mixture } s_f = 1.8604 \text{ kJ/kg.K}$$

$$s_{fg} = 4.9603 \text{ kJ/kg.K}$$

$$x_7 = \frac{s_7 - s_f}{s_{fg}}$$

$$= \frac{6.4219 - 1.8604}{4.9603} = 0.9196$$

$$h_7 = h_f + x_7 h_{fg} = 640.09 + 0.9196 * 2108.0$$

$$h_7 = 2578.6 \text{ kJ/kg}$$

Turbine exit:

$$p_8 = 20 \text{ kPa} \, h_f$$

$$= 251.42 \text{ kJ/kg}$$

$$s_8 = s_6 = 6.4219 \text{ kJ/kg.k} \quad h_{fg}$$

$$= 2357.5 \text{ kJ.kg}$$

$$\text{Mixture } s_f = 0.832000 \text{ kJ.kg}^{-1}.\text{K}^{-1}$$

$$s_{g.f} = 7.05200 \text{ kJ.kg}^{-1}.\text{K}^{-1}$$

$$x_8 = \frac{s_8 - s_f}{s_{fg}} = \frac{6.4219 - 0.8320}{7.0752}$$

$$x_7 = 0.7901$$

$$h_8 = h_f + x_8 h_{fg}$$

$$= 251.42 + 0.7901 \times 2357.5$$

$$h_8 = 2114.0 \text{ kJ/kg}$$

When the entire steam is routed to the process heater:

$$W_T = 5(h_6 - h_7)$$

$$W_T = 5(3242.4 - 2578.6)$$

$$W_T = 3319 \text{ kW}$$

$$W_P = 5(h_7 - h_3)$$

$$W_P = 5(2578.6 - 640.09)$$

$$W_P = 9693 \text{ kW} \quad \textbf{Ans.}$$

(b)

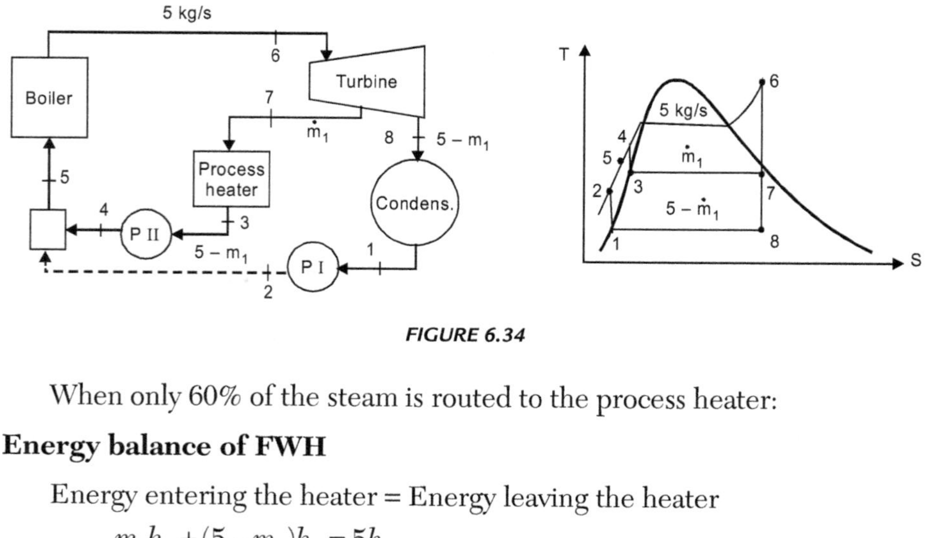

FIGURE 6.34

When only 60% of the steam is routed to the process heater:

Energy balance of FWH

Energy entering the heater = Energy leaving the heater

$$m_1 h_4 + (5 - m_1) h_2 = 5 h_5$$

$$3 \times 650.47 + 2 \times 650.47 = 5 \times h_5$$

$$h_5 = 494.91 \text{ kJ/kg}$$

$$W_T = 5(h_6 - h_7) + 2(h_7 - h_8)$$

$$W_T = 5(3242.4 - 2578.6) + 2(2578.6 - 2114.0)$$
$$W_T = 3319 \text{ kW}$$
$$W_P = 3(h_7 - h_3)$$
$$W_P = 3(2578.6 - 640.09)$$
$$W_P = 5816 \text{ kW} \quad \textbf{Ans.}$$

6.6.8 Binary Vapor Cycle

Binary vapor cycle uses two fluids. Figure 6.35 shows parts of a binary vapor power plant.

1. **Mercury cycle:** Mercury boiler, mercury turbine, mercury condenser or steam generator, and mercury feed pump.

2. **Steam cycle:** Steam generator or mercury condenser, steam super heater, steam turbine, steam condenser, and water feed pump.

A. Thermodynamic properties of mercury:

Mercury has its following thermodynamic properties:

1. The boiling pressure of 12.5 bar only is necessary to match the steam temperature of 540°C. High pressure is not necessary to obtain high main temperature of heat supply.

2. Mercury does not have any erosive or corrosive effects on the metal parts.

3. The boiling point is 354.4°C and the freezing point is 3.3°C at atmospheric pressure.

4. Mercury can be safely used with high-temperature metals.

5. The saturation state of liquid can be obtained by vertical expansion almost isentropic as for Carnot cycle.

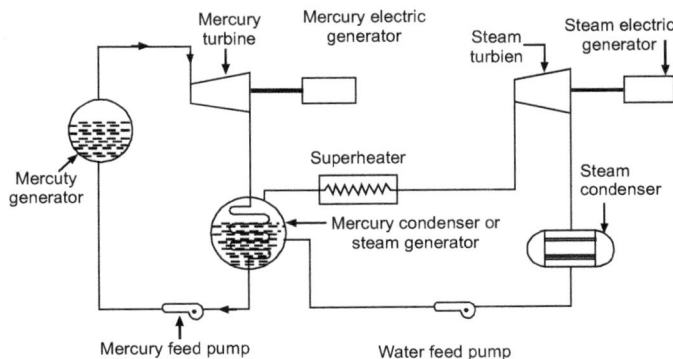

FIGURE 6.35 Elements of binary vapor power plant.

B. Binary vapor cycle analysis:

w_s = Work done per kilogram of steam

h_s = Heat supplied per kilogram of steam produced in the mercury condenser or steam boiler

η_s = Thermal efficiency of steam cycle.

m = Mass of mercury in the Hg cycle per kilogram of steam produced in the steam cycle.

η_{hg} = Thermal efficiency of Hg cycle.

H_{hg_1} = Heat required for Hg per kilogram of steam in mercury boiler.

W_{hg} = Work produced per kilogram of Hg cycle

h_{hg_2} = Heat removed from Hg per kilogram of steam in the mercury condenser.

Work done in mercury cycle per kilogram of water flow:

$$W_{hg} \cdot m = w \tag{6.62}$$

Work done in steam cycle per kilogram of water flow:

$$W_s \cdot 1 = w \tag{6.63}$$

Heat supplied in Hg boiler per kilogram of water flow:

$$h_{hg_1} \cdot m = h_t \tag{6.64}$$

Overall work done in binary vapor cycle:

$$W_t = W_s + W_{hg} \cdot m \tag{6.65}$$

$\therefore$ Total efficiency of binary vapor cycle:

$$\eta = \frac{\text{Work done}}{\text{Heat supplied}}$$

$$= \frac{W_t}{h_t} = \frac{mW_{hg} + W_s}{mh_{hg_1}} \tag{6.66}$$

Thermal efficiency of Hg cycle:

$$\eta_{hg} = \frac{mW_{hg}}{mh_{hg_1}}$$

$$= \frac{W_{hg}}{h_{hg_1}} = \frac{h_{hg_1} - h_{hg_2}}{h_{hg_1}} = 1 - \frac{h_{hg_2}}{h_{hg_1}} \tag{6.67}$$

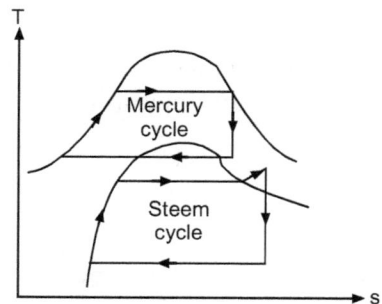

FIGURE 6.36 Binary vapor cycle on T-S graph.

$$= \frac{mh_{hg_1} - h_8}{mh_{hg_1}} = 1 - \frac{1}{m} \cdot \frac{h_8}{h_{hg_1}} \tag{6.68}$$

Heat supplied by the Hg vapor = Heat received by the steam

$$mh_{hg_2} = h_s \cdot 1 \tag{6.69}$$

But

$$\eta_{hg} = 1 - \frac{h_{hg_2}}{h_{hg_1}} \tag{6.70}$$

Thermal efficiency of steam cycle:

$$\eta_s = \frac{W_s}{h_s} = \frac{h_s - h_{s_2}}{mh_{hg_2}} \tag{6.71}$$

Combining overall thermal effectiveness, the following was obtained:

$$\eta_{\text{cycle}} = \eta_{hg} + h_s(1 - \eta_{hg}) \tag{6.72}$$

There is a problem in the design of binary cycle plants. Special plant design is required to avoid leakage of dangerous Hg vapor. Cost of Hg inventory is very high. Pumping of Hg and heat transfer are difficult.

6.7 STIRLING CYCLE

The Stirling cycle is similar to Carnot cycle which was invented in 1815 by Dr. Robert Stirling. Figure 6.37 shows an ideal Stirling cycle on *T-S* and *P-V* diagrams and consists of the following four thermodynamic processes:

1. 1–2: reversible isothermal compression.

2. 2–3: heat addition by reversible isochoric process.

3. 3–4: reversible isothermal expansion.

4. 4–1: reversible isochoric heat rejection.

The working material for Stirling cycle is a gas such as hydrogen, air, etc. and not water vapor. Similar to Carnot cycle, all processes in a perfect Stirling cycle are reversible. Heat pump or refrigerator works on reversed Stirling cycle. This cycle has the application for cryogenic engine.

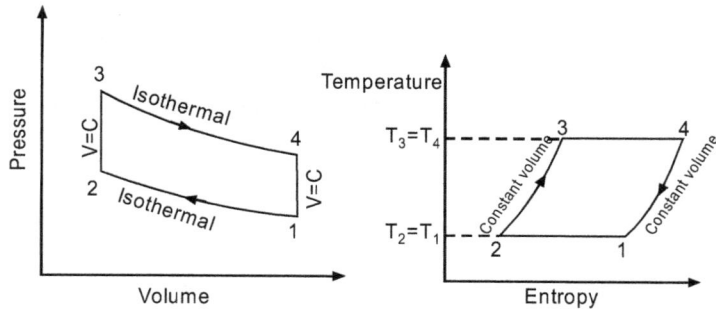

FIGURE 6.37 Stirling cycle.

6.8 ERICSSON CYCLE

The Ericsson cycle is another perfect cycle of thermodynamics discovered by John Ericsson. *T-S* and *P-V* diagrams for ideal Ericsson cycle are shown in Figure 6.38. It consists of the following processes:

1. 1–2: reversible isothermal compression process.

2. 2–3: heat addition by reversible isobaric process.

3. 3–4: reversible isothermal expansion process.

4. 4–1: reversible isobaric heat rejection process.

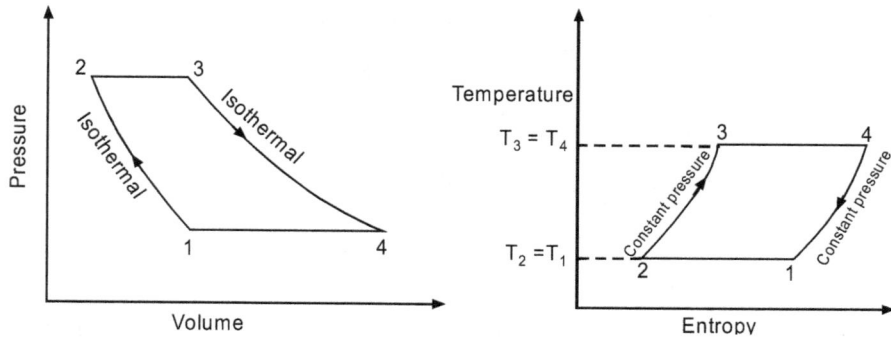

FIGURE 6.38 Ericsson cycle.

This cycle is not used in engines but are used in gas turbine plants.

Stirling, Ericsson, and Carnot cycles are identical in thermodynamics. In theory, each cycle can obtain same maximum efficiency. None of these cycles have found applications in practical engineering.

6.9 REFRIGERATION CYCLE

In vapor refrigeration cycles, the heat is moved from lower temperature to higher temperature at the cost of external work. Figure 6.39 shows the mechanical components of a refrigeration cycle when used as air conditioner.

The COP is defined as follows:

$$(COP)_R = \frac{Q_L}{W_C}, \tag{6.73}$$

where Q_L is the useful heat energy extracted from refrigerated space, and W_C is the work energy consumed by the compressor. The reverse process is used as a basic for heat pump cycle that extracts heat from the ambient air in winter and heat is transferred to the indoor space. COP of a heat pump is expressed as follows:

$$(COP)_{HP} = \frac{Q_H}{W_C} \tag{6.74}$$

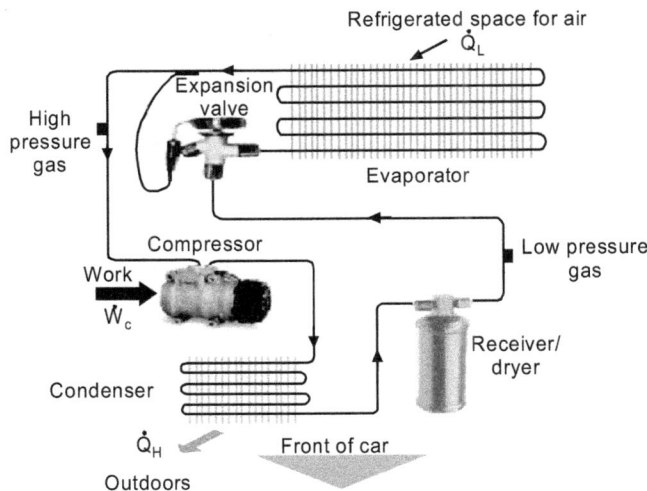

FIGURE 6.39 Refrigeration cycle.

EXAMPLE 6.8

One kilogram of air contained in a piston–cylinder apparatus occupies a 0.4700 m^3·kg^{-1} of specific volume at 250°C and executes a cycle between two heat reservoirs which are, respectively, at 250°C and 150°C temperatures. As shown in Figure 6.40, the cycle proceeds as follows: (1–2): Expanding isothermally to specific volume of 1.19 m^3·kg. (2–3):

Cooling to 150°C at a constant volume. (3–4): Isothermal compression to specific volume of 0.47 m³·kg. Heating to 250°C at constant volume. Show that equation $\oint \delta Q = \oint \delta W$ holds for this cycle. Calculate the thermal efficiency of the cycle.

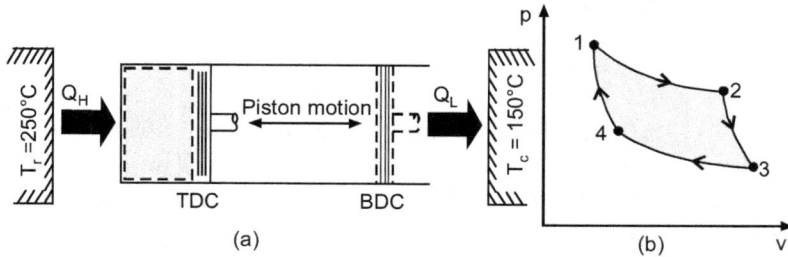

FIGURE 6.40 Schematic diagram of Example 6.8.

Solution: The net heat transfer to the system may be evaluated as follows:

$$Q_{net} = Q_{41} + Q_{12} + Q_{23} + Q_{34}$$

From Figure 6.40(b),

$$T_1 = T_2, \ T_3 = T_4, \ V_1 = V_4, \ V_2 = V_3,$$

$$\begin{aligned} Q_{41} &= U_1 - U_4 \\ &= mc_v(T_1 - T_4) \\ &= 1 \times 0.718 \times (523 - 423) = 71.8 \text{ kJ}, \end{aligned}$$

$$\begin{aligned} Q_{12} = W_{12} &= m_R T_1 \ \text{In} \ \frac{V_2}{V_1} \\ &= 1 \times 0.718 \times 523 \times \text{In} \frac{1.19}{0.49} = 139.44 \text{ kJ} \end{aligned}$$

$$\begin{aligned} Q_{23} &= U_3 - U_2 = mc_v(T_3 - T_2) \\ &= -1 \times 0.718 \times (423 - 523) = -71.8 \text{ kJ}, \end{aligned}$$

$$\begin{aligned} Q_{34} &= m_R T_3 \ \text{In} \ \frac{V_4}{V_3} \\ &= 1 \times 0.718 \times 423 \times \text{In} \frac{0.47}{1.19} = -112.78 \text{ kJ} \end{aligned}$$

Net heat transported through this cycle is calculated as follows:

$$Q_{net} = 71.8 + 139.44 + (-71.8) + (-112.78) = 26.66 \text{ kJ}.$$

Net work transported by the cycle can be expressed as follows:

$$W_{net} = W_{14} + W_{12} + W_{23} + W_{34}$$

The processes (4–1) and (2–3) are similar. Work terms W_{41} and W_{23} are zero. Also,

$$W_{12} = m_R T_1 \text{ In} \frac{V_2}{V_1}$$

$$= 1 \times 0.718 \times 523 \times \text{In} \frac{1.19}{0.47} = 139.44 \text{ kJ}$$

and,

$$W_{34} = m_R T_3 \text{ In} \frac{V_4}{V_3}$$

$$W_{34} = 1 \times 0.718 \times 423 \times \text{In} \frac{0.47}{1.19} = -112.78 \text{ kJ}$$

$$W_{net} = 0 + 139.44 + 0 + (-112.78) = 26.66 \text{ kJ},$$

$$\therefore \quad Q_{net} = W_{net},$$

$$\therefore \quad Q_H = Q_{41} + Q_{12} = 71.8 + 139.44 = 211.24 \text{ kJ}.$$

where Q_H is the heat received from the heat reservoir.

$$W_{net} = 26.66 \text{ kJ.kg}^{-1}$$

$$\eta_{th} = \frac{26.66}{211.44} \times 100\%$$

$$\eta_{th} = 12.62\% \quad \textbf{Ans.}$$

REVIEW QUESTIONS

1. Write a brief note on the types of thermodynamic cycles.

2. With the help of a diagram, explain the main elements of a heat cycle. Also define the main performance parameters.

3. With the help of suitable diagram, describe the Carnot Vapor Power Cycle. Also explain h_{Carnot} and limitations of the cycle.

4. Explain in detail about the principle of Carnot cycle as applicable to heat engine and refrigerator.

5. What is Reverse Carnot cycle?

6. Describe in detail about the ideal Rankine cycle. Give its comparison with the Carnot cycle.

7. With the help of T-S diagram, explain the effects of the following operating conditions on the performance of Rankine cycle.

   a. Lowering the condenser pressure.

   b. Raising the boiler pressure.

   c. Superheating the steam.

8. What is regeneration? Discuss the various types of regeneration system.

9. Describe a regeneration system with the following:

   a. Open FWHs.

   b. Closed FWHs.

10. Differentiate between the following:

    a. Closed FWH system versus drains cascaded backward.

    b. Closed FWH system versus drains pumped forward.

11. What is reheating? With the help of suitable diagrams, discuss a reheat cycle.

12. What is cogeneration? With the help of a diagram discuss a cogeneration system.

13. Describe an Hg-steam binary vapor power cycle. What are the advantages and limitations?

14. Write notes on the following:

    a. Stirling cycle.

    b. Ericsson cycle.

15. Explain the performance of a refrigeration cycle suitable for air conditioning.

NUMERICAL EXERCISES

1. Dry saturated steam is supplied to a steam turbine at 5 MPa. Condenser pressure is 5 kPa. Show the Rankine cycle on T-S diagram and determine the simple Rankine efficiency.   **(36.67%)**

2. A steam power plant working on Rankine cycle has steam supply pressure of 20 bar and condenser pressure of 0.5 bar. If the initial condition of steam is dry and saturated, calculate the Carnot and Rankine efficiencies of the cycle neglecting the pump work. **(27%, 33%)**

3. A steam power plant working on Rankine cycle has a steam pressure of 100 bar and 550°C and condenser pressure of 0.05 bar at the turbine's inlet. Determine the cycle's efficiency, SSC, and work ratio, where all processes are reversible. **(43%, 2.55 kg/kWh, 0.99)**

4. A regenerative Ericsson hot air engine works between temperature limits of 318 K and 503 K. The expansion ratio is 2. Calculate the following:

   i. Work done per kilogram of air.
   ii. Cycle efficiency. **(36.76 kJ/kg, 36.8%)**

3. A 5-tonne refrigeration plant is working on R-12 with condenser and evaporator temperatures as 40°C and –10°C, respectively. Determine the following:

   i. The refrigerant flow rate in kg/s. **(0.18 kg/s)**
   ii. The volumetric flow rate of compressor in m^3/s. **(0.0139 m^3/s)**
   iii. The compressor's outlet temperature. **(48°C)**
   iv. The heat rejection in the condenser in kW. **(24.27 kW)**
   v. The pressure ratio. **(4.39)**
   vi. The quality of refrigerant after throttling. **(30.5%)**
   vii. The COP of plant. **(4.14)**
   viii. The power required to drive the compressor. **(4.72 kW)**
   ix. Compare the COP with that of a Carnot refrigerator working between 40°C and –10°C. **(0.787)**

THERMODYNAMIC APPLICATIONS

7.1 ENGINEERING APPLICATIONS OF STEADY FLOW ENERGY EQUATION

For a system at steady-state conditions, the energy remains constant with time. In general, the energy conservation principle for such system gives the following:

$$\sum \dot{m}_e \left(h + \frac{1}{2}v^2 + gz \right)_e = \sum \dot{m}_i \left(h + \frac{1}{2}v^2 + gz \right)_i , \qquad (7.1)$$

where $\dot{m}_i$ and $\dot{m}_e$ are mass flow rates of incoming and outgoing fluids. Therefore, for solving Equation (7.1) for a system, the application of the principle of mass conservation is important.

$$\therefore \quad \sum \dot{m}_i = \sum \dot{m}_e$$

The total mass flow at the inlet of the system is equal to the total mass flow at the outlet of the system.

Energy analysis of the important engineering equipment will be discussed. The conservation of mass and energy for the two-port systems is as follows:

$$\dot{m}_i = \dot{m}_e \qquad (7.2)$$

$$\left(h + \frac{1}{2}V^2 + gz \right)_e - \left(h + \frac{1}{2}V^2 + gz \right)_i = q - w, \qquad (7.3)$$

where q and w are the specific heat and work transfer of the system.

7.2   DIFFUSERS AND NOZZLES

Diffusers and nozzles are devices used for changing the velocity of a flowing stream. In steam or gas turbine applications, a nozzle increases the velocity of a fluid at the expense of a pressure drop in the direction of flow. A diffuser, however, increases the fluid pressure in the direction of flow at the expense of a decrease in velocity. In centrifugal compressors, the increase in pressure of an accelerated fluid is provided by such devices.

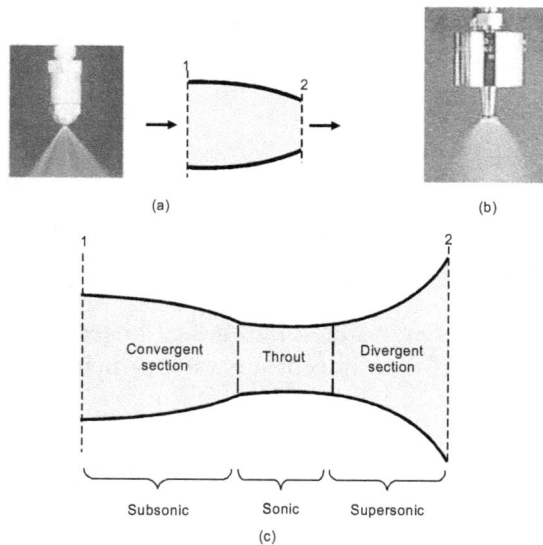

FIGURE 7.1 (a) Converging nozzle, (b) diverging nozzle, and (c) converging–diverging nozzle.

Under subsonic or supersonic flow conditions, the general shapes of a nozzle or a diffuser are shown in Figure 7.1. A diffuser for the supersonic flow or a nozzle for the supersonic flow must have a decreasing cross-sectional area in the direction of flow as shown in Figure 7.1(a). For subsonic flows, however, the opposite occurs. In Figure 7.1(b), the spray nozzle, which is used for the humidification processes, is a supersonic nozzle with an increasing cross-sectional area along the direction of flow. If a fluid has to be accelerated from a subsonic to a supersonic velocity, a converging–diverging nozzle, as shown in Figure 7.1(c), must be used. In such applications, the fluid assumes the sonic flow conditions are at the throat.

Since both of these devices are essentially ducts, no shaft work is involved, and for most conditions, due to the high velocity of the fluid, the heat transfer through the walls is negligible. The potential energy change may also be neglected. Owing to single inlet and outlet, Equations (7.2) and (7.3) are applicable to such devices. With these assumptions, Equation (7.3) may be simplified as follows:

$$\left(h + \frac{1}{2}V^2\right)_e = \left(h + \frac{1}{2}V^2\right)_i \quad (7.4)$$

By rearranging, the velocity of a fluid at any cross section of a nozzle or a diffuser may be calculated as follows:

$$V_e = \sqrt{V_i^2 + 2(h_i - h_e)} \quad (7.5)$$

In the case of a nozzle, when the pressure and the enthalpy of the fluid are reduced together, acceleration of the fluid flow is accrued. The pressure of steam is changed to its kinetic energy supplied to the turbine.

EXAMPLE 7.1

Steam enters a nozzle with a stagnation enthalpy of 2780 kJ/kg and mass flow rate of 9 kg/min at 1. At the nozzle's exit, the steam has a velocity of 1070 m/s and a specific volume of 18.75 m³/kg. Determine, at the nozzle's exit,

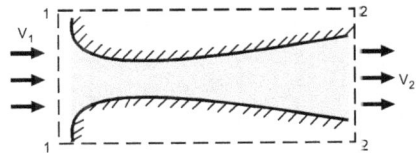

FIGURE 7.2 Schematic of Example 7.1.

a. the enthalpy of the steam,
b. the cross-sectional area of the nozzle.

Solution:
a. A stagnation enthalpy of a fluid is the summation of the kinetic energy and the enthalpy by h.

$$h_0 = h_1 + \frac{1}{2}V_1^2$$

$$h_0 = 2780 \text{ kJ/kg}$$

$$h_2 = h_0 - \frac{1}{2}V_2^2$$

$$h_2 = 2780 - \frac{1070^2}{2}$$

$$h_2 = 2207.5 \text{ kJ/kg} \quad \textbf{Ans.}$$

b. For one-dimensional flow, considering the continuity of equation, the cross-sectional area of the nozzle at the exit can be determined as follows:

$$A_2 = \frac{\dot{m}v_2^2}{2}$$

$$A_2 = \frac{0.15.18.75}{1070}$$

$$A_2 = 0.00262 \text{ m}^2 \quad \textbf{Ans.}$$

7.3 WATER TURBINE

Figure 7.1 shows a water turbine, receiving water from the height. The potential energy of water is converted into kinetic energy when it enters the turbine, and part of it is transformed into useful work that is used to produce electricity. Taking turbine shaft center as datum, the equation of energy can be written as follows:

$$\left(u_1 + p_1 v_1 + z_1 g + \frac{C_1^2}{2}\right) + Q = \left(u_2 + p_2 v_2 + z_2 g + \frac{C_2^2}{2}\right) + W \tag{7.6}$$

Now,
$$Q = 0.0$$
$$\Delta u = -u_1 + u_2 = 0$$
$$v_1 = v_2 = v$$
$$z_2 = 0$$

$$\therefore \quad \left(p_1 v + z_1 g + \frac{C_1^2}{2}\right) = \left(p_2 v + z_2 g + \frac{C_2^2}{2}\right) + W \tag{7.7}$$

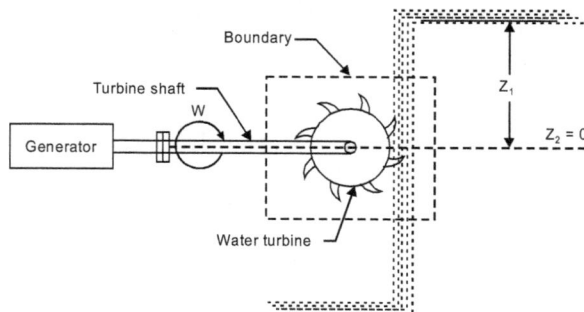

FIGURE 7.3 Water turbine.

Work is done by the system, therefore W is positive.

7.4 THROTTLING DEVICES

A throttling device is used to reduce the pressure of the flow by an obstruction. An orifice, a valve, a long capillary tube, and a porous plug are some of the throttling devices.

A reduction in pressure by an orifice in Figure 7.4(a) is used to measure the velocity of a fluid flowing through a pipe. In Figure 7.4(b), by reducing the cross-sectional area of the flow by a valve, a greater pressure drop across the valve occurs and the flow rate decreases. The throttling devices are widely used in refrigeration units, and are also utilized in reducing the power of an engine.

In a throttling process, no work interaction is involved, and usually the transport of heat of the process is neglected. Furthermore, the changes of kinetic and potential energies can be neglected. Thus, for a throttling process, Equation (7.8) is simplified as follows:

$$h_i = h_e \tag{7.8}$$

By which it is stated that a throttling process can be considered as a constant enthalpy process.

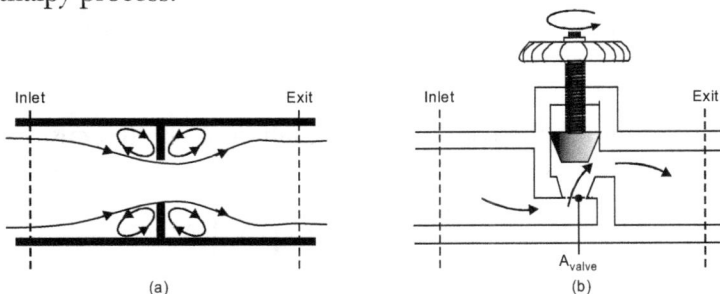

FIGURE 7.4 Throttling processes.

EXAMPLE 7.2

High pressure water at the temperature of 200°C and 20 bar in pressure is adiabatically throttled to 15 bar by a valve (Figure 7.5). Determine:

a. The exit temperature and the quality of water.

b. The velocity if $A_2 = 2A_1$, and $V_1 = 10$ m/s.

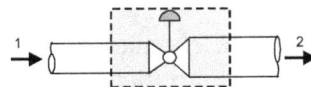

FIGURE 7.5 Throttling devices.

Solution:

a. The enthalpy of compressed water and the enthalpy of saturated water may be assumed to be same at 200°C. From steam tables, take $h_1 = 852.45$ kJ/kg.

$$\therefore \quad h_2 = h_1 = 852.45 \text{ kJ/kg.}$$

For saturated steam at $p_2 = 15$ bar, $h_{2f} < h_2 < h_{2g}$. Therefore, water at the exit of the valve is a mixture of liquid and vapor. Thus, the exit temperature becomes $T_2 = 198.3°C$ and the quality of water is calculated as follows:

$$x_2 = \frac{h_2 - h_{2f}}{h_{2g}}$$

$$x_2 = \frac{825.45 - 844.84}{1947.3}$$

$$x_2 = 0.0039 \quad \textbf{Ans.}$$

b. The specific volume at the exit of the valve is calculated as follows:

$$v_2 = v_{2f} + x_2 v_{2fg}$$
$$= 0.00115 + 0.0039 \times 0.1306$$
$$= 0.00166 \text{ m}^3\text{/kg.}$$

Considering the continuity of equation of one-dimensional flow, the exit velocity becomes

$$v_2 = \left(\frac{A_1}{A_2}\right)\left(\frac{v_2}{v_1}\right)V_1$$

$$v_2 = \left(\frac{1}{2}\right)\left(\frac{0.00166}{0.001}\right) \times 10$$

$$v_2 = 8.3 \text{ m/s} \quad \textbf{Ans.}$$

7.5 MIXING CHAMBERS

In chemical industry, certain fluids at different temperatures are mixed in appropriate ratios of mass so that a desired temperature is obtained. In a central air-conditioning system, outdoor air is mixed with cooled air for providing desired conditions in the indoor space. The mixing of fluids in such devices must be at the same pressure. There is no work transfer through such devices, and the chambers are usually insulated. The alteration in potential energy and kinetic energy of fluids entering and leaving the chamber may be neglected. Accordingly, the continuity and the energy equations become,

$$\sum \dot{m}_i = \sum \dot{m}_e \qquad (7.9)$$

$$\sum \dot{m}_i h_i = \sum \dot{m}_e h_e \qquad (7.10)$$

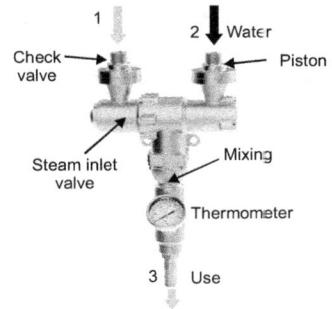

FIGURE 7.6 Steam–water mixing device.

Figure 7.6 shows a steam–water mixer which provides low-pressure hot water by using cold water and steam.

Water and steam valves are fixed at the entrance, and the water temperature is measured by a thermometer at the exit.

EXAMPLE 7.3

Hot water is supplied for a process by mixing steam at 10 bar and 0.2 in quality to the mixing chamber (Figure 7.6) at a rate of 1 kg/s. After throttling to 5 bar, the steam is mixed with 2 kg/s of water at 30°C. Measure the mass flow rate and the temperature of the mixture at the chamber's outlet.

Solution: With respect to the boundary around the chamber, Equations (7.9) and (7.10) become,

$$\dot{m}_1 + \dot{m}_2 = \dot{m}_3$$
$$\dot{m}_1 h_1 + \dot{m}_2 h_2 = \dot{m}_3 h_3$$

By using the steam tables,

$$h_1 = h_{1f} + x_1 h_{1fg}$$
$$= 762.81 + 0.2 \times 2015.3 = 843.42 \text{ kJ/kg,}$$
$$h_2 = 125.79 \text{ kJ/kg}$$

The enthalpy at the chamber's exit will be $h_3 = 365$ kJ/kg. Since, $h_3 < h_{3f}$ at $p_3 = 5$ bar, the water at the chamber's outlet is a compressed liquid, and the temperature is

$$T_3 = 87.3°C.$$

7.6 GAS TURBINE OR STEAM TURBINE

In a gas or steam turbine, gas or steam is passed into the turbine and portion of its energy is converted into work. The turbine runs a generator for production of electricity as shown in Figure 7.7. The gas or steam leaves the turbine at a low temperature and pressure.

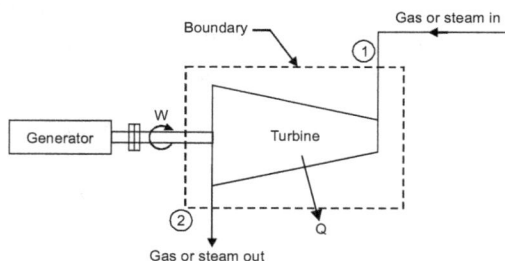

FIGURE 7.7 Steam or gas turbine.

These machines are used in a large number of processes frequently encountered in industry varying from power protection to refrigeration, and from liquid transportation in pipe lines to air handling duct systems.

The energy content of the working fluid is converted into mechanical shaft work. As the fluid does not work on the rotating blades of this machine, the fluid's temperature and pressure decrease in the direction of flow. A single-stage, axial flow

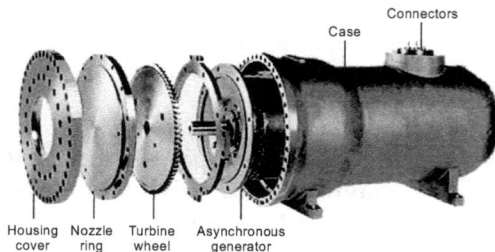

FIGURE 7.8 Single-stage steam turbine.

steam turbine is presented in Figure 7.8. As shown in Figure 7.8, the steam is first accelerated by the nozzles, and rotates the blades that are attached to the turbine wheel. Rotating shaft in turn turns the generator. The change in the potential energy between the inlet and the outlet of a turbine is usually neglected. Unless stated, the heat losses through the turbine casing are small in comparison to the enthalpy changes. Therefore, for a turbine, Equation (7.3) is simplified as follows:

$$W = \left(h + \frac{1}{2}V^2\right)_i - \left(h + \frac{1}{2}V^2\right)_e \tag{7.11}$$

7.7 COMPRESSORS AND PUMPS

Compressors and pumps are utilized in compressing or raising the pressure of a fluid. A compressor that uses a gas or a vapor as the working fluid can be either rotating or reciprocating type. In a rotary compressor, the fluid is accelerated and the kinetic energy is increased by the shaft work done on the impeller. Then, in the diffuser section, deceleration of fluid causes an increase in fluid's pressure. Figure 7.9(a) displays a rotary compressor which might be used for raising the pressure of the inlet air steam of a jet engine. In a household refrigerator, however, a reciprocating compressor is used to increase the refrigerant's vapor pressure. A sectional view of a pump is shown in Figure 7.9(a). Pumps are used to raise the pressure of liquids.

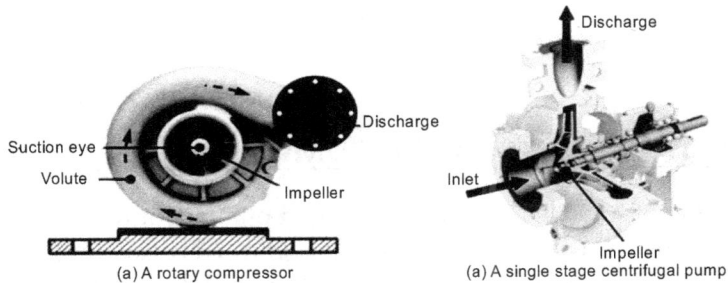

(a) A rotary compressor   (a) A single stage centrifugal pump

FIGURE 7.9 Single-stage rotary compressor and pump.

Both machines consume work energy. Again, the change in the potential energy between the inlet and the outlet of a compressor is negligible. For compressors with a large power capacity either fins are attached to the compressor body or water is circulated through the jackets of the compressor's cylinders for cooling. Hence, in general, the energy equation for a compressor might be written as follows:

$$W = \left(h + \frac{1}{2}V^2 \right)_i - \left(h + \frac{1}{2}V^2 \right)_e \tag{7.12}$$

7.8 FANS AND BLOWERS

Fans are almost universally used for circulation of air or other gases through low-pressure systems. As shown in Figure 7.10, a centrifugal fan is widely used for moving large or small quantities of air over an extended range of pressures at the expense of work consumption. The heat transfer through the housing of a fan and the change in the potential energy of the fluid are usually neglected. Therefore, Equation (7.3) is simplified as follows:

$$W = \left(h + \frac{1}{2}V^2 \right)_i - \left(h + \frac{1}{2}V^2 \right)_e \tag{7.13}$$

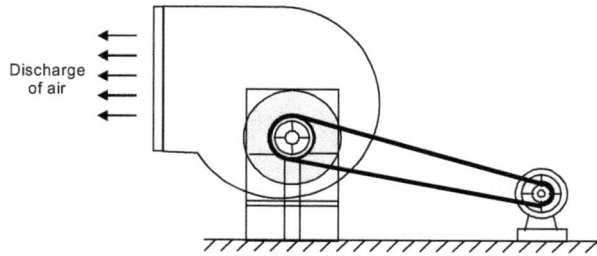

FIGURE 7.10 Centrifugal fan.

7.9 CENTRIFUGAL WATER PUMP

Pulling of water from a lower level and pumping it to higher level is done by centrifugal water pump as shown in Figure 7.11. Running the pump requires work, which may be provided by a diesel engine or electric motor.

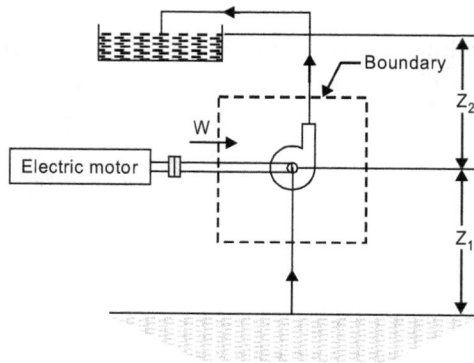

FIGURE 7.11 Centrifugal water pump.

Here $\Delta u = 0$ and $Q = 0$ because there is no change in the water temperature.

$$v_1 = v_2 = v$$

The equation of energy becomes:

$$p_1 v_1 + z_1 g + \frac{C_1^2}{2} = p_2 v_2 + z_2 g + \frac{C_2^2}{2} - W \qquad (7.14)$$

Work is applied on the system so the signal of W is negative.

7.10 CENTRIFUGAL COMPRESSOR

Figure 7.12 shows a centrifugal compressor that is used to supply air at moderate pressure and large quantity.

$$\Delta Z = 0$$

Work W is supplied to the system, therefore it is taken as negative and also heat Q is taken as negative as it is lost from the system.

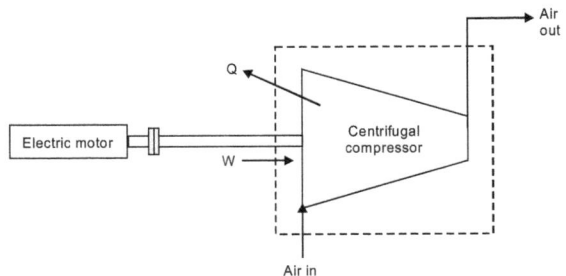

FIGURE 7.12 Centrifugal compressor.

The energy equation becomes:

$$\left(h_1 + \frac{C_1^2}{2}\right) - Q = \left(h_2 + \frac{C_2^2}{2}\right) - W \qquad (7.15)$$

EXAMPLE 7.4

A small steam turbine working at the partial load produces 100 kW of power at the flow rate of 0.3 kg/s. The steam (Figure 7.13) is throttled before entering the turbine from 1.5 MPa at 300°C to 1 MPa pressure. For an exhaust pressure of 10 kPa, calculate:

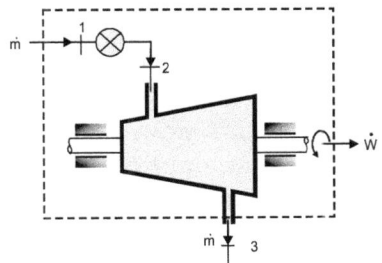

FIGURE 7.13 Schematic of Example 7.4.

a. The steam state at the outlet of the turbine.

b. The exhaust duct diameter for a steam velocity of 20 m/s.

Solution:

a. In Equation (7.13), neglecting the change in the kinetic energy, the enthalpy at the outlet becomes $h_3 = h_1 - w$, where $h_1 = 3037.6$ kJ/kg, and the turbine specific power is:

$$W = \frac{\dot{W}}{\dot{m}} = \frac{100}{0.3} = 333.3 \text{ kJ/kg}$$

Thus, the enthalpy relation yields $h_3 = 2704.3$ kJ/kg at $p_3 = 10$ kPa.

Using the superheated steam tables, the temperature and the specific volume at the turbine's outlet are as follows:

$$T_3 = 108.8°C, \ v_3 = 17.6 \ m^3/kg \quad \textbf{Ans.}$$

b. Assuming one-dimensional flow in the exhaust duct,

$$\dot{m}C_3 = A_3 V_3$$

$$A_3 = \frac{\dot{m}C_3}{V_3}$$

$$= \frac{\pi d_2^3}{4}$$

$$= \frac{0.3 \times 17.6}{20} = 0.264 \ m^2$$

And the duct $d_3 = 0.58$ m **Ans.**

EXAMPLE 7.5

An axial flow compressor in Figure 7.14 intakes air at 100 kPa and 27°C, and compresses to 500 kPa and 227°C with an outlet velocity of 100 m/s. For a shaft power of 50 kW, determine the amount of air flowing though the compressor in one hour.

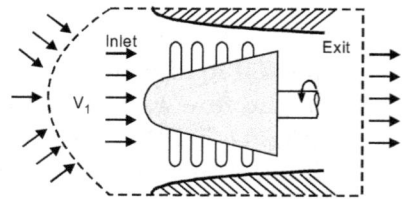

FIGURE 7.14 Schematic of Example 7.5.

Solution: Assuming the compression process to be adiabatic, Equation (7.11) is simplified as follows:

$$W = \left(\frac{V^2}{2} + h \right)_1 - \left(\frac{V^2}{2} + h \right)_2$$

Since the atmospheric air is stagnant, $V_1 = 0$.

And
$$h_1 - h_2 = c_p(T_1 - T_2)$$
$$= 1.005 \times (300 - 500)$$
$$= -206 \ kJ/kg.$$

$$W = -201 - \frac{100^2}{2000} = -206 \ kJ/kg$$

Specific shaft work w as

$$w = \frac{\dot{W}}{\dot{m}}$$

The mass flow rate becomes

$$\dot{m} = \frac{\dot{W}}{w} = \frac{50}{206}$$

$$\dot{m} = 0.424 \text{ kg/s } \textbf{Ans.}$$

The amount of air flowing through the compressor in one hour will be 871.2 kg. **Ans.**

7.11 RECIPROCATING COMPRESSOR

Figure 7.15 shows a reciprocating compressor which supplies the air at a high pressure in small quantities (comparing with centrifugal compressor). The reciprocating compressor can be considered as a steady flow system by using a receiver which decreases the flow instabilities completely.

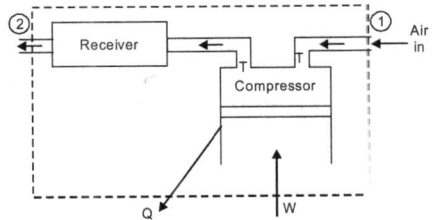

FIGURE 7.15 Reciprocating compressor.

Applying the equation of energy to the process,

$$\Delta KE = 0.0$$

and $$\Delta PE = 0.0$$

$$h_1 - h_2 = Q - W \qquad (7.16)$$

∴

EXAMPLE 7.6

A reciprocating compressor (Figure 7.16) is cooled by water circulated through the water jacket of the compressor. The compressor intakes the air at 1 bar and 25°C and compresses to 7 bar and 80°C. The cooling water circulating at a rate of 0.3 kg/s enters at 15°C and leaves at 35°C. If the volumetric rate of flow for the air at the inlet is 0.05 m³/s, determine the shaft power.

Solution:

The mass flow rate of supplied air:

$$\dot{m}_a = \rho_i \cdot \dot{V}_i,$$

where

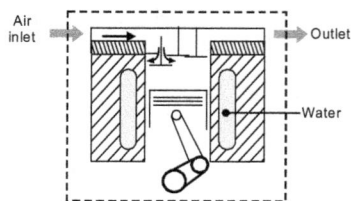

FIGURE 7.16 Reciprocating compressor cooling.

$$\rho_i = \frac{P_i}{RT_i}$$

$$= \frac{100}{0.287 \times 298} = 1.169 \text{ kg/m}^3$$

And

$$\dot{m}_a = 1.169 \times 0.05$$
$$\dot{m}_a = 0.058 \text{ kg/s}$$

Neglecting the change in the kinetic energy of air,

$$W = q + (h)_i - (h)_e$$

And the amount of heat removed by water is calculated as follows:

$$\dot{Q}_w = \dot{m}_w c_w (T_2 - T_1)$$
$$\dot{Q}_w = 0.3 \times 4.18 \times (35 - 15)$$
$$\dot{Q}_w = 25.08 \text{ kW}$$

The heat loss of air during the compression process equals the heat removed by water.

Then, $\dot{Q}_a = -\dot{Q}_w = -25.08$ kW,

And the transfer of heat per unit mass of air is calculated as follows:

$$q = \frac{\dot{Q}_a}{\dot{m}_a}$$

$$= \frac{-25.08}{0.058} = -432.41 \text{ kJ/kg}$$

$$h_1 - h_2 = c_p (T_i - T_e)$$
$$= 1.005 \times (25 - 80)$$
$$= -55.275 \text{ kJ/kg}.$$

From equation of energy, substituting these values

$$w = -432.41 + (-55.275)$$
$$= -487.685 \text{ kJ/kg.}$$

The shaft power of the compressor is calculated as follows:

$$\dot{W} = \dot{m}w = 0.058 \times -487.685$$

Or $\qquad \dot{W} = -28.285 \text{ kW}$ **Ans.**

7.12 BOILER

A boiler can be used to transform heat to the entering water and produce the steam. The boiler system is shown in Figure 7.17.
For this system,

$$\Delta \left(\frac{C_2^2}{2} \right) = 0.0 \text{ and } \Delta Z = 0.0.$$

Because there is no work is created and not absorbed so W is equal to 0.0
The system's equation of energy becomes

$$h_2 = Q + h_1 \tag{7.17}$$

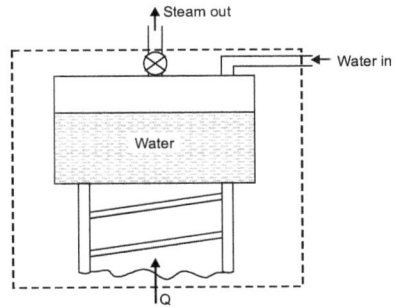

FIGURE 7.17 Boiler.

7.13 THE HEAT EXCHANGERS

A heat exchanger is an apparatus which supplies heat energy to two or more fluids at various temperatures. Transfer of heat between the fluids happens by the separation wall.

The fluids do not mix because they are parted by a surface of heat transfer. Common examples of heat exchangers are the tube and shell exchangers, dry cooling towers, evaporators, automobile radiators, condensers, and air preheaters.

The heat exchanger in Figures 7.18 and 7.19 is composed of active heat exchanging sections as a matrix including the passive fluid distribution sections as baffles, headers, seals, entering and exiting nozzles, and the heat transfer surface. There are no moving parts in a heat exchanger.

The heat loss to the environment from a heat exchanger is usually small in comparison to the heat transfer between the two fluids. There is no work transfer and the changes of potential and kinetic energies of both fluids can be neglected.

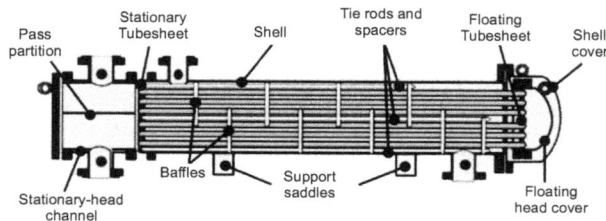

FIGURE 7.18 Elements of heat exchanger.

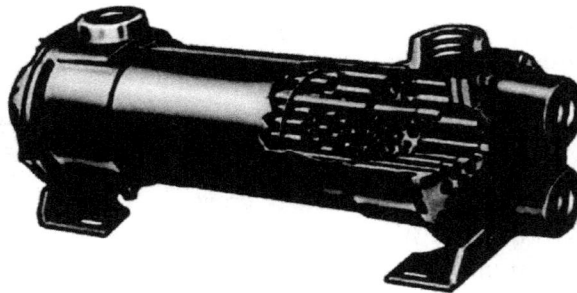

FIGURE 7.19 Cut front of heat exchanger.

Equation (7.1) is simplified as follows:

$$\dot{m}_c(h_e - h_i)_c = \dot{m}_h(h_i - h_e)_h \tag{7.18}$$

7.14 CONDENSER

A condenser is used to condense the steam in steam power plants and the vapor of refrigerant in the refrigeration system by the use of air or water as a cooling medium. A condenser is shown in Figure 7.20.

For this system, $\Delta KE = 0.0$ and $\Delta PE = 0.0$.

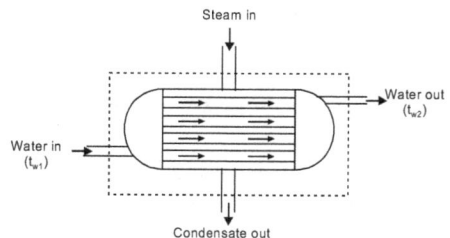

FIGURE 7.20 Condenser.

$W = 0.0$ (no work is created or absorbed).
Using the equation of energy for steam flow,

$$Q = h_1 - h_2 \qquad (7.19)$$

One kilogram of steam flowing through the condenser has heat Q.

The transfer of heat between water and steam is the only transfer in the system.

Heat absorbed by the cooling water is calculated as follows:

$$Q = m_w c_w (t_{w2} - t_{w1}) = m_w (h_{w2} - h_{w1})$$

From Equation (7.8),

$$Q = m_w c_w (t_{w2} - t_{w1}) = m_w (h_{w2} - h_{w1}) = h_1 - h_2, \qquad (7.20)$$

where m_w = mass flow rate of cooling water and c_w = specific heat of water.

7.15 EVAPORATOR

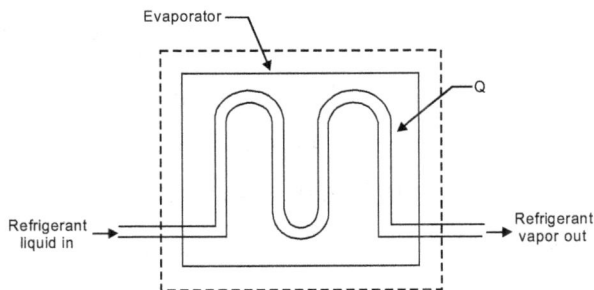

FIGURE 7.21 Evaporator.

For this system, $\Delta KE = 0.0$ and $\Delta PE = 0.0$

Work is neither absorbed nor provided, $W = 0$

By using the equation of energy to the system, the following is obtained.

$$h_2 - h_1 = Q \qquad (7.21)$$

The system's temperature is less than the temperature of environment. Heat flows into the system, and therefore Q is taken as positive.

EXAMPLE 7.7

Figure 7.22(a) shows a layout of a room's air conditioner in which refrigerant R22 is used as the working fluid. The warm room air at

100 kPa, 27°C enters the evaporator at a flow rate 6m³/min. As shown in Figure 7.22(b), the refrigerant at 245 kPa pressure and 0.3 quality flows into the tubes of the evaporator at a rate of 1 kg/min, and at the same pressure leaves as saturated vapor. Calculate the temperature of air at the evaporator's outlet.

Solution:

The hot and the cold fluids are, respectively, the room air and the refrigerant. The mass flow rate of air is calculated as follows:

$$\dot{m}_a = \rho_i \cdot \dot{V}_i,$$

where

$$\rho_i = \frac{P_i}{RT_i}$$

$$= \frac{100}{0.287 \times 300} = 1.161 \text{ kg/m}^3$$

And

$$\dot{m}_a = 1.161 \times \frac{6}{60} = 0.116 \text{ kg/s}$$

The enthalpy change of the refrigerant is calculated as follows:

$$(h_e - h_i)_c = (1 - x_1) \cdot h_{fg}$$

$$0.7 \times 220.33 = 154.231 \text{ kJ/kg}$$

From Equation (7.18),

$$\dot{m}_c (h_e - h_i)_c = \dot{m}_h (h_i - h_e)_h$$

$$0.0116 \times 154.231 = 0.0116 \times 1.005 \times (27 - T_e)$$

And solving for T_e, the temperature of air at the evaporator's exit becomes

$$T_e = 5.03°C \textbf{ Ans.}$$

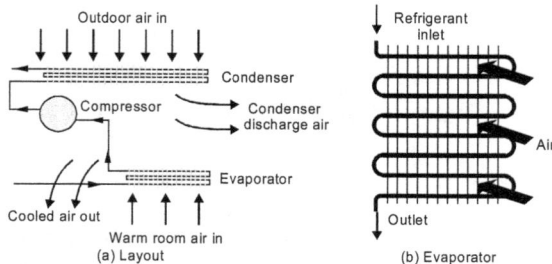

FIGURE 7.22 Layout of a room's air conditioner.

7.16 FLOW IN DUCTS AND PIPES

There is no need to signify the engineering importance of transporting a liquid or a gas between two stations. The pressure drop happens when fluids flow through a duct or a restraining channel. This drop of pressure is called fluid friction. The magnitude of fluid friction is influenced by different factors such as fluid density, velocity, viscosity, condition of its surface, shape or diameter of duct part, types of flow turbulent or viscous, and fluid temperature and pressure.

A piping or a duct system has only one inlet and outlet. Hence, the energy equation in the form of Equation (7.3) is applicable to such systems. In certain applications, in addition to flow friction, heat transfer takes place. There are numerous examples for such a case like flow of water through the tubes of a boiler, flow in pipes of a heat exchanger, or in flow of a refrigerant through the tubes of an evaporator or a condenser. In these applications, transfer of heat cannot be neglected in the energy equation. For a flow of air in distribution ducts of an air-conditioning system, however, the walls of the channel are insulated and the system is taken to the adiabatic condition. In cases of installing a fan, a pump, or an electric heater to a pipe system, then the work transfer in Equation (7.3) has to be retained.

Let us consider a flow of an incompressible fluid in an insulated duct. The enthalpy change for an incompressible fluid is calculated as follows:

$$\Delta h = \Delta u + v\Delta p, \text{ and } \Delta u = c\ \Delta T \qquad (7.22)$$

Due to fluid friction, certain amount of heat will be produced at the conduit walls. Because of insulation, this heat is regained as an increase in the fluid's internal energy. However, suppose the fluid is to be non-viscous, then there is no friction, and the process of flow becomes isothermal. Thus, for an incompressible, non-viscous flow of a fluid through an insulated duct together with Equation (7.22), Equation (7.23) is simplified as follows:

$$w = \left[v(P_2 - P_1) + \frac{1}{2}(V_2^2 - V_1^2) + g(z_2 - Z_1) \right] \qquad (7.23)$$

This is known as the Bernoulli's equation.

EXAMPLE 7.8

A hair dryer is basically a duct in which a 100-W electric resistor and a 50-W fan are installed. The fan steadily sucks in the ambient air at 100 kPa, 20°C and forces it over a resistor such that the exit temperature of air is 50°C. Calculate:

a. Velocity of the air at the exit.

b. Volumetric flow rate of the air at the entrance.

The exit of dryer has a cross-sectional area of 20 cm².

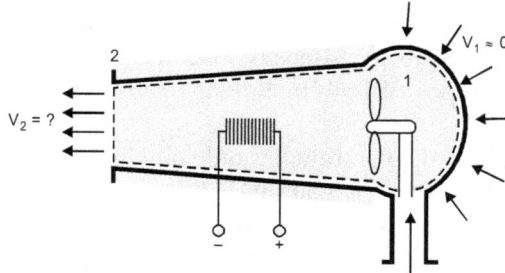

FIGURE 7.23 Schematic of Example 7.8.

Solution:

a. The mass flow rate is calculated as follows:

$$\dot{m} = \rho_2 V_2 A_2,$$

where

$$\rho_2 = \frac{P_2}{RT_2} = \frac{100}{0.287 \times 300} = 1.078 \text{ kg/m}^3$$

And

$$\dot{m} = 0.00215 \ V_2$$

The enthalpy change of the refrigerant is calculated as follows:

$$(h_2 - h_1) = c_p(T_2 - T_1)$$
$$= 1.005 \times (50 - 20)$$
$$= 30.15 \text{ kJ/kg}$$

The total work transferred to the duct is calculated as follows:

$$\dot{W} = \dot{W}_e + \dot{W}_f$$
$$= -0.1 + (-0.05) = -0.15 \text{ kW}$$

There is no change in the potential energy of this flow, and the surface of the duct is assumed to be insulated. Therefore, Equation (7.3) becomes

$$30.15 + \frac{V_2^2}{2} = \frac{-0.15}{0.00215 \, V_2}$$

Where the velocity of ambient air is taken to be zero, $V_1 = 0$. Rearranging this expression resulted as follows:

$$V_2^3 + 60.29 \, V_2 - 139.534 = 0$$

Which yields

$$V_2 = 2.147 \text{ m/s.}$$

b. The mass flow rate is calculated as follows:

$$\dot{m} = 0.00215 \, V_2 = 0.00125 \times 2.147 = 0.0046 \text{ kg/s.}$$

With respect to the inlet conditions,

$$\dot{m} = \rho_1 \dot{V}_1$$

$$\rho_1 = \frac{P_1}{RT_1} = \frac{100}{0.287 \times 293} = 1.189 \text{ kg/m}^3$$

And the volumetric flow rate at the inlet is calculated as follows

$$\dot{V}_1 = \frac{\dot{m}}{\rho_1}$$

$$\dot{V}_1 = \frac{0.0046}{1.189} = 0.00386 \text{ m}^3\text{/s} \quad \textbf{Ans.}$$

7.17 UNSTEADY FLOW PROCESSES

In engineering practice, the unsteady flow processes are as common as the steady flow processes. In a transient flow process, the average mass flow and energy inside and outside of the control volume are not same.

The process of charging of compressed gas tanks is not a steady flow process because during this process, fluid mass flow rate as well as the temperature and the pressure of the fluid in the tank will vary with time. The system is called transient system. One of the applications of the transient systems is to run a turbine by the pressurized air of a storage tank, for fulfilling the energy needs of a facility during a cutoff period.

In transient systems, the shape and the boundary might vary with time, and the boundary can be in motion with respect to a reference point. Besides, an accurate calculation of the terms is required for mass and energy equation. However, analytic method can still be devised to get results reasonable for a system under study. Thus, to get a solution, the following simplifications can be used for transient system:

1. The thermodynamic state of every point within the boundary of the system is the same at a given time. The properties of the system are single valued. However, the state of the system is time variant.

2. The state of the fluid flowing into the system is time invariant. This assumption can be fulfilled by extending the boundary of the system to a region where the fluid properties do not vary with time. The flow rate of incoming fluid, however, might be a function of time.

3. The fluid flowing out of the system is assumed to be at the same stage as the system. This assumption requires that the fluid is essentially in equilibrium at all times. Hence, a discharge of a fluid must be slow enough that the process is quasistatic.

In regard to these assumptions, the kinetic energy change is neglected. Unless indicated, the change in the potential energy is usually negligible.

Two subsequent cases will be studied:

1. Filling of a tank.

2. Draining of a tank.

7.17.1 Filling of a Tank

Assume,

T_1 = initial temperature of fluid
T_2 = final temperature of fluid
v_1 = initial specific volume of fluid
v_2 = final specific volume of fluid
p_1 = initial pressure of fluid
p_2 = final pressure of fluid
m_1 = initial mass of fluid
m_2 = final mass of fluid
u_1 = initial specific internal energy of fluid
u_2 = final specific internal energy of fluid

C' = inlet velocity of fluid
v' = inlet specific volume of fluid
T' = inlet temperature of fluid
P' = inlet pressure of fluid
h' = inlet specific enthalpy of fluid
u' = inlet specific internal energy of fluid

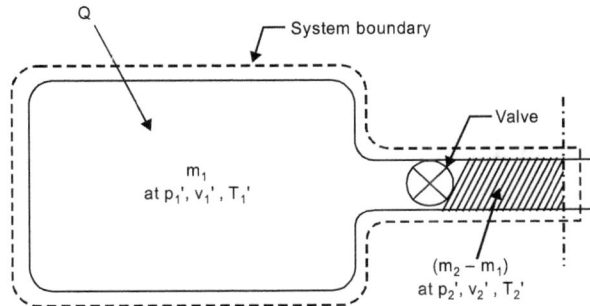

FIGURE 7.24 Filling of a tank.

The amount of fluid entering is calculated as follows:

$$= -(m_1 - m_2)$$

The energy of the fluid entering is calculated as follows:

$$= (m_2 - m_1)\left(u' + p'v' + \frac{C'^2}{2} \right) \tag{7.24}$$

$$= (m_2 - m_1)\left(h' + \frac{C'^2}{2} \right) \tag{7.25}$$

Heat transported to the control volume is calculated as follows:

$$Q = (m_2 - m_1)\left(h' + \frac{C'^2}{2} \right) + Q = m_2 u_2 - m_1 u_1 \tag{7.26}$$

If the tank is completely insulated, there is no transfer of heat.

$$Q = 0.0$$

And

$$Q = (m_2 - m_1)\left(h' + \frac{C'^2}{2} \right) = m_2 u_2 - m_1 u_1 \tag{7.27}$$

Initially, if the tank is empty and completely insulated,

$$m_1 = 0.0$$

Therefore,

$$h' + \frac{C'^2}{2} = u_2 \qquad (7.28)$$

Similarly, by neglecting the kinetic energy in the pipe line, the following is obtained.

$$u_2 = h' \qquad (7.29)$$

7.17.2 Draining of a Tank

Similar to the tank's filing process, the energy equation can be taken as follows:

$$(m_1 - m_2)\left(h' + \frac{C'^2}{2}\right) - Q = m_1 u_1 - m_2 u_2 \qquad (7.30)$$

where C' = outlet velocity of fluid and h' = outlet specific enthalpy of fluid.

For a completely drained tank and no heat transfer and negligible exit velocity,

$$u_1 = h' \qquad (7.31)$$

7.18 EXERGY EFFICIENCY OF HEAT EXCHANGER

The fluids do not mix because they are separated by a heat transfer surface. For adiabatic condition of heat exchanger and steady state, exergy equation will be as follows:

$$\dot{m}_h(\phi_1 - \phi_2) = \dot{m}_c(\phi_4 - \phi_3)_c + \dot{I}$$

Decrease in the exergy of the hot fluid = increase in the exergy of the cold fluid

Exergy destruction = 0

The exergy efficiency of a heat exchanger is calculated as follows:

$$\eta_{hx} = \frac{\dot{m}_c(\psi_4 - \psi_3)_c}{\dot{m}_h(\psi_1 - \psi_2)_h} \qquad (7.32)$$

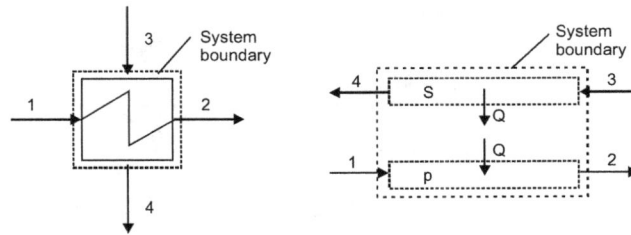

FIGURE 7.25 Exergy efficiency of heat exchanger.

As

$$\eta_{Ex,u} = \frac{\sum Ex_{out}}{\sum Ex_{in}} = \frac{Ex_2 + Ex_4}{Ex_1 + Ex_3}$$

$$\sum Ex_{out} = \sum Ex_{in} - \sum Ex_{loss}$$

$$Ex_2 + Ex_4 = Ex_1 + Ex_3 - Ex_{loss}$$

Thus

$$\eta_{Ex,u} = \frac{Ex_1 + Ex_3 - Ex_{loss}}{Ex_1 + Ex_3}$$

Corrected definition of heat exchanger efficiency:

$$\eta_{Ex,f} = \frac{\sum Ex_{product}}{\sum Ex_{source}} = \frac{Ex_2 - Ex_1}{Ex_3 - Ex_4}$$

$$\Rightarrow \qquad \eta_{Ex,f} = \frac{Ex_3 - Ex_4 - Ex_{loss}}{Ex_3 - Ex_4}$$

EXAMPLE 7.9

An air heater has tubes of 10 cm diameter and water enters the tubes at a mass flow rate of 24 kg/min at 140°C and 0.5 MPa and exits at 60°C and 0.5 MPa. Air is the cold fluid that enters heat exchanger at volumetric flow rate of 100 m³/min. The inlet conditions are 25°C and 110 kPa, and the used air velocity is 25 m/s. At the outlet, the pressure of air is 110 kPa. If the surrounding is at $P_0 = 100$ kPa, $T_0 = 25$°C, determine:

a. Outlet temperature of the air.
b. Efficiency of the heat exchanger.

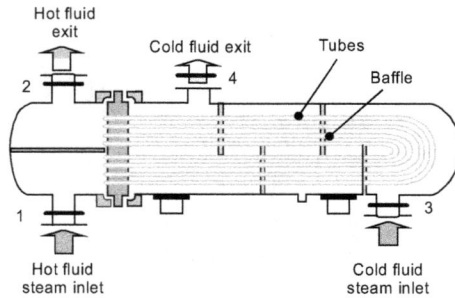

FIGURE 7.26 Heat exchanger.

Solution:

a. At steady-state condition, the equation of energy is as follows:

$$\dot{m}_w(-h_2 + h_1) = \dot{m}_a c_{pa}(T_4 - T_3)$$

Assuming air as an ideal gas,

$$\rho_3 = 1 / \left(\frac{RT_3}{P_3} \right)$$

$$= \frac{110.00}{0.287 \times 298} = 1.286 \text{ kg/m}^3$$

$$\dot{m}_a = \rho_3 V_3$$

$$\dot{m}_a = 2.14700 \text{ kg/s}$$

Using thermodynamic properties of water,

$$(T_4 - T_3) = \frac{0.4(589.1300 - 251.1300)}{2.14700 \times 1.005}$$

$$(T_4 - T_3) = 62.6500 \text{ K}$$

$$T_4 = 87.6500°\text{C} \quad \textbf{Ans.}$$

b. By neglecting the change in the velocity of water, the change of exergy of water stream is calculated as follows:

$$\dot{m}_w(\psi_1 - \psi_2) = 0.4 \times [(589.13 - 251.13) - 298 \times (1.7391 - 0.8312)]$$

$$\dot{m}_w(\psi_1 - \psi_2) = 26.97800 \text{ kW}$$

The cross-sectional area of the air tubes at the inlet is assumed to be same as at the outlet.

$$\rho_3 V_3 = \rho_4 V_4$$

$$V_4 = \frac{T_4}{T_3} V_3$$

$$V_4 = 1.21 \times 25 = 30.25 \text{ m/s}$$

The exergy change of air stream is calculated as follows:

$$\dot{m}_a (\psi_4 - \psi_3) = 2.147 \times \left[\begin{array}{c} 1.005 \times (87.65 - 25) - 298 \times 1.005 \\ \times \ln\left(\dfrac{360.65}{298}\right) + \dfrac{1}{2000}(30.25^2 - 25^2) \end{array} \right]$$

$$= 12.843 \text{ kW}$$

The heat exchanger's efficiency, from Equation (7.32), is calculated as follows:

$$\eta_{hx} = \frac{12.843}{26.978} = 0.476$$

$$\eta_{hx} = 47.6\% \quad \textbf{Ans.}$$

It shows that 52.4% of the useful energy has been wasted at the inlet of heat exchanger, but from exact efficiency of energy, there is no waste, and energy efficiency is 100%. This is not correct as total transfer of useful energy from hot fluid to cold fluid cannot happen.

7.19 EXERGY EFFICIENCY OF EXPANSION (TURBINE) AND COMPRESSION (COMPRESSOR)

Turbines and compressors transform fluid energy. Assuming that there is no heat loss to the environment and the system operates at steady condition, energy recovery by a turbine can be determined by using the following equation:

$$\underbrace{\dot{W}_t}_{\text{Shaft work}} = \underbrace{\dot{m}(\phi_1 - \phi_2)}_{\text{Decrease in the exergy supplied}} - \underbrace{\dot{I}}_{\text{Exergy destruction}} \tag{7.33}$$

Exergy supplied to a turbine is $\dot{m}_w (\psi_1 - \psi_2)$.

The exergy efficiency is calculated as follows:

$$\eta_{tx} = \frac{\dot{W}_t / \dot{m}}{\psi_1 - \psi_2} = \frac{w}{w_{\text{rev}}} \tag{7.34}$$

$\Psi_1 - \Psi_2$ is the maximum useful work, that is, W_{rev}. This work can be extracted out from the stream of fluid at similar inlet and outlet conditions. Equation (7.34) gives two dissimilar definitions of exergetic efficiency of a turbine. A pump, fan, or compressor with no transfer of heat to the surrounding and at steady condition, taking, $-X'w = W'c$ in Equation (7.35), the exergy equation will be as follows:

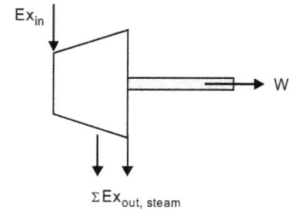

FIGURE 7.27 Steam turbine efficiency.

$$\underbrace{\dot{m}(\phi_1 - \phi_2)}_{\text{Increase in the exergy of the provide}} = \underbrace{\dot{W}_c}_{\text{Shaft power}} - \underbrace{\dot{I}}_{\text{Exergy destruction}} \qquad (7.35)$$

The percentage increase in the exergy of flow to the provided exergy represents the exergy efficiency. For a fan or a compressor at adiabatic condition,

$$\eta_{cx} = \frac{\Psi_2 - \Psi_1}{\dot{W}_c / \dot{m}} = \frac{w_{rev}}{w_c}$$

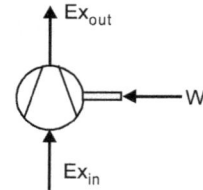

FIGURE 7.28 Compressor.

Exergy efficiency of steam turbines is shown in Figure 7.27.

$$\eta_{cx} = \frac{Ex_{product}}{Ex_{source}} = \frac{w}{Ex_{in} - \sum Ex_{out, steam}}$$

Exergy efficiency of compressors is shown in Figure 7.28.

EXAMPLE 7.10

Steam entering an adiabatic turbine with a velocity of 80 m/s at 600°C and 6 MPa and leaving at 100°C and 50 kPa is shown in Figure 7.28. The speed at the outlet is 140 m/s. The conditions of surrounding are 25°C and 100 kPa, and the shaft power of turbine is determined to be 5 MW. Calculate the exergy loss in the turbine.

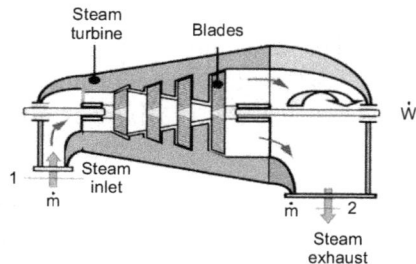

FIGURE 7.29 Steam turbine.

Solution:

The exergy equation can be simplified as follows:

$$\Phi_2 - \Phi_1 = (-U_1 + U_2) + P_0(-V_1 + V_2) + T_0(S_1 - S_2)$$

In the case of Example 7.10,

$$\dot{X}_w = 5.00 \text{ MW}, \dot{X}_Q = 0.0,$$

For a steady-state flow, the shaft work will be

$$\dot{W} = \dot{m}\left[(-h_2 + h_1) + \frac{1}{2}(V_1^2 - V_2^2)\right] = \dot{X}_w$$

The exergy relation is given as follows:

$$\dot{I} = \dot{m} \, T_0(s_1 - s_2)$$

The mass flow rate is calculated as follows:

$$\dot{m} = 5.1500 \text{ kg/s.}$$

For

$$\Delta s = 0.52700 \text{ kJ/kgK}$$

$$T_0 = 298.00 \text{ K,}$$

The wastage of exergy is calculated as follows:

$$\dot{X}_{\text{destroyed}} = 810.04 \text{ kW.}$$

7.19.1 Exergy Efficiency of Combustion

Exergy efficiency of adiabatic combustor is shown in Figure 7.30.

$$\eta_{ex} = \frac{Ex_{\text{product}}}{Ex_{\text{source}}} = \frac{Ex_{\text{flue gas}}^{tm} - Ex_{\text{flue}}^{tm} - Ex_{\text{air}}^{tm}}{Ex_{\text{flue}}^{ch} + Ex_{\text{air}}^{ch} + Ex_{\text{flue gas}}^{ch}}$$

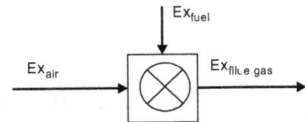

FIGURE 7.30 Adiabatic combustor.

With: $Ex_{\text{flue gas}} = Ex_{\text{flue gas}}^{tm} + Ex_{\text{flue gas}}^{ch}$ etc.

$$Ex^{tm} = \text{thermo} - \text{mechanical energy}$$

$$Ex^{ch} = \text{chemical energy}$$

If fuel and oxidant are supplied at environmental temperature,

$$Ex_{\text{air}}^{tm} = 0 \text{ and } Ex_{\text{flue}}^{tm} = 0$$

And the equation is simplified as follows:

$$\eta_{ex} = \frac{Ex_{\text{flue gas}}^{tm}}{Ex_{\text{flue}}^{ch} + Ex_{\text{air}}^{ch} - Ex_{\text{flue gas}}^{ch}}$$

EXAMPLE 7.11

The burning gas flows at a speed of 80 m/s through the turbo-jet engine nozzle at 747°C and 260 kPa, and leaves at 5000°C and 70 kPa. The ambient temperature is 17°C. Calculate:

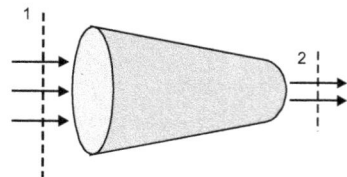

FIGURE 7.31 Flow through nozzle.

a. The speed of gas at the nozzle's exit.
b. The exergy rate of reduction of the gas at a mass flow rate of 0.1 kg/s.

Suppose $k = 1.300$ and $c_p = 1.1500$ kJ/kg·K of the burning gas.

Solution:

a. For this specific condition, there is neither work nor heat exchange across the nozzle's boundary. The equation of energy at a steady-state condition is given as follows:

$$h_2 + \frac{V_2^2}{2} = h_1 + \frac{V_1^2}{2}$$

Using the given data, the velocity of gas leaving the nozzle is calculated as follows:

$$V_2 = 757.900 \text{ m/s } \textbf{Ans.}$$

b. The equation of exergy is given as follows:

$$\dot{I} = \left[(-h_2 + h_1) + \frac{1}{2}(V_1^2 - V_2^2) \right] \dot{m} + T_0 \, \dot{m}(s_2 - s_1)$$

Then
$$\dot{I} = -\dot{m} + T_0 \, \dot{m}(s_2 - s_1)$$

Suppose if the gas behaves as an ideal gas, the change of entropy can be determined as follows:

$$R = \frac{k-1}{k} c_p$$

$$s_2 - s_1 = \int_1^2 C \frac{dT}{T}$$

The exergy rate reduction due to mixing and friction of fluid atoms is given as follows:

$$\dot{I} = 0.84100 \text{ kW } \textbf{Ans.}$$

REVIEW QUESTIONS

1. What are diffusions and nozzles? What are their types and uses?

2. Derive an energy equation for a water turbine.

3. Write notes on:

   a. Throttling devices

   b. Mixing chambers

   c. Gas and steam turbines

   d. Compressors and pumps

   e. Fans and Blowers

4. Derive energy equation for the following:

   a. Centrifugal water pump

   b. Centrifugal air compressor

   c. Reciprocating compressor

5. Write notes on heat exchangers such as boilers, condenser, and evaporators.

6. Write Bernoulli's equation for flow through ducts and pipes.

7. Explain the energy equations for the following unsteady flow processes:

   a. Filling a tank

   b. Emptying a tank.

8. Describe the exergy efficiency of heat exchange, expansion, compression, and combustion processes.

NUMERICAL EXERCISES

1. In an isentropic flow through a nozzle, air flows at the rate of 600 kg/hr. At the nozzle's inlet, pressure is 2 MPa and temperature is 127°C. The exit pressure is 0.5 MPa. If the initial air velocity is 300 m/s, determine:

    a. The exit velocity of air.

    b. The inlet and exit area of nozzle. **(594 m/s, 31.88 mm^2, 43.35 mm^2)**

2. Water vapor at 90 kPa and 150°C enters a subsonic diffuser with a velocity of 150 m/s and leaves the diffuser at 190 kPa with a velocity of 55m/s. During this process, 1.5 kJ/kg of heat is lost to the surrounding. Determine:

    a. The final temperature.

    b. The mass flow rate.

    c. The exit diameter assuming the inlet diameter as 10 cm and steady flow. **(154°C, 0.543 kg/s, 11.42 cm)**

3. Following are the details of a steam turbine:

Steam flow rate	= 1 kg/sec
Inlet velocity of steam	= 100 m/sec
Exit velocity of steam	= 150 m/sec
Enthalpy at inlet	= 2900 kJ/kg
Enthalpy at outlet	= 1600 kJ/kg

    Write the steam flow energy equation. Assuming that the change in the potential energy is negligible, determine the power available from the turbine. **(1293.75 kW)**

4. An air compressor compresses atmospheric air at 0.1 MPa and 27°C by ten times of inlet pressure. During compression, the heat loss to the surrounding is estimated to be 5% of compression work. Air enters the compressor with a velocity of 40 m/s and leaves with a velocity of 100 m/s. Inlet and exit from cross-sectional areas are 100 cm^2 and 20 cm^2, respectively. Estimate the temperature of air at exit from compressor and power input to the compressor. **(1498 K, 5467.86 kW)**

CHAPTER 8

MEASUREMENT THEORY AND INSTRUMENTS

8.1 INTRODUCTION

A measurement system is the one which connects the observer with the process. A typical measurement system is depicted in Figure 8.1. The measured variable is considered to be the information variable. The observer is offered with the values of the information variable. The true value of the variable is the input to measurement system and the measured value of the variable is the output of measurement system.

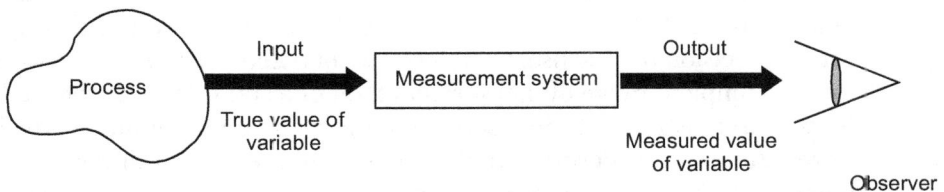

FIGURE 8.1 Measurement system

8.2 PERFORMANCE AND PURPOSE OF MEASUREMENT SYSTEMS

A process is known as a system, which is used to generate information. Examples are a jet fighter, a chemical reactor, a weather system, the human heart, a car, a submarine, a gas platform, etc.

Acceleration, velocity, and displacement variables are generated by a car and the composition, pressure, and temperature variables are generated by a chemical reactor.

In a perfect measurement system, the measured value should be the same as the true value. The precision of the system means the nearness between the true value and the measured value. A complete precise system is a theoretical system. A precision real system is quantified.

The error of measurement system (E) can be defined by the following equation:

$$E = \text{(measured value)} - \text{(true value)}$$
$$E = \text{(system output)} - \text{(system input)}$$

If the measured value of the gas flow rate inside your pipe is 11.0 m³/h while the true value is 11.2 m³/h, the error will be $E = -0.2$ m³/h. If the measured value of the engine rotational velocity is 3140 rpm and the true value is 3133 rpm, $E = +7$ rpm. This is the main performance index for a measurement system.

8.3 MEASUREMENT VARIABLES

8.3.1 Calibration

Calibration is a comparison between the sensor or instrument output under test versus the instrument output of known precision when the same value of measurement input is used for both instruments.

This process is carried out for a range of inputs, which cover the full range of measurement of the sensor or instrument. Calibration assures that the measuring precision of all sensors and instruments used in a measurement system for the complete range of measurement, on one condition that the calibrated sensors and instruments are used in the same environment under which they were calibrated. If sensors and instruments are used under different environments, suitable rectifications have to be done. Calibration steps used for a single sensor will be applicable to all instruments.

8.3.2 Units of Measurement

The units of measurement were used in exchange trading to determine the sums to be paid and to make rules of the relative values of various merchandises. Early measuring systems were dependent on whatsoever was available as a measuring unit. The human torso was a suitable example for the objective of measuring the length and it granted the units of the cubit, the hand, and the foot. Although mostly enough for swap commerce systems, these units of measurement are inaccurate, changing from one body to another. There has been a progressive action towards units of measurement which must be more precise.

Many units were named for measuring the same physical variable. For example, length can be measured in meters, yards, etc. In addition to main length units, sub-division in units is available such as inches, feet, millimeters, and centimeters, with a relation between these units. For example, 1 feet is equal to 12 inches, 1 mile is equal to 1760 yards, and 1 yard is 3 feet. A metric system is now used which contains, for example, the meter unit and its sub-divisions (millimeter and centimeter) for measuring the value of length.

8.3.3 Measurement Uncertainty

The instrument accuracy means closeness of the instrument reading output to the true value. It is more customary to quote the inaccuracy value of measurement instead of the instrument precision value.

Inaccuracy of measurement is the range in which reading maybe erroneous and is quoted as a proportion of the full-scale reading of a device.

The maximum error of measurement is attached to the full-scale reading of the instrument. Measured values are less than the full-scale reading. This means that the potential error of measurement is magnified. It is important that devices are selected for a range, which will cover the values to be measured. If the pressure has to be measured between 0 and 1 bar, do not use the instrument for range of measurement between 0 and 10 bar.

EXAMPLE 8.1

The range of measurement of a pressure gauge is 0–10 bar and a quoted inaccuracy is ±1.0% on full-scale reading.

1. **Find the maximum error obtained in measurement for this instrument.**

2. **Find the probable measurement error obtained as a proportion of the output reading if the pressure gauge is used to measure air pressure value of 1 bar.**

Solution:

1. The maximum error obtained in all measurement reading is 1.0% of the full-scale reading, which is 10 bar for this device. Thus, the maximum probable error is $1.0\% \times 10$ bar $= 0.1$ bar

2. The maximum error of measurement is a fixed value attached to full-scale reading of instrument, regardless of the amount that the device is really measuring. The maximum error value is 0.1 bar. Therefore, when measuring the pressure of 1 bar, the maximum error of this pressure is 10% of the measured value, i.e., 0.01 bar.

8.3.4 Measurement Sensitivity and Repeatability

The measurement sensitivity is defined as a measure of the variation in output of instrument when the amount being measured varies by a given quantity.

$$\text{Sensitivity} = \frac{\text{Deflection scale}}{\text{Measure and producing deflection}}$$

The measurement sensitivity is the slope of the straight line.

Repeatability shows the proximity of output readings when the same value of the input is applied frequently through a short time period, with the same device and observer, same conditions of measurement, same conditions of utilizing maintained all the time, and the same location.

8.4 MEASUREMENT OF TEMPERATURE

Temperature is the property of a system that indicates the potential for heat transfer with other systems. Therefore, two systems are said to have equal temperature when there is no heat transfer between them.

Temperature can be measured by a thermometer based on the molecular or internal energy of the body.

The first measurement of temperature was executed by Galileo at the end of 16th century. His thermometer was based on air expansion. A scale was connected to his device to point out "heat degrees." It is necessary to have unanimous measurement units. For temperature, the degree of Celsius (C°) and the Kelvin for the international units.

8.4.1 Concept of Temperature

Temperature measures the change of kinetic energy with temperature in a material. Kinetic energy is the movement of molecules: the molecules of solids vibrate in place, but molecules of liquids and gases move freely. All the molecules do not move with the same speed. With higher temperature, the molecules move faster. The temperature is defined as a measure of potential of stored energy in a mass of matter.

Temperature is the potential used to produce heat transmit from a higher temperature point to a lower temperature point. The temperature does not measure the amount of heat in a matter because molecules have both potential and kinetic energy. Temperature can only measure kinetic energy and not potential energy. The heat transfer rate is a function of temperature change.

8.4.2 Temperature Scales

To compare and measure temperatures, a temperature scale should be available. These temperature scales are defined in an expression of physical properties that take place at fixed temperatures.

i.   The Celsius temperature (°C)

The Celsius temperature scale is known by international convention in the expression of two fixed points, the steam under the ice points. The ice point temperature is defined as 0°C and the steam point temperature as 100°C. The ice point is a temperature of water and ice at a pressure of 1.013×10^5 N.m^{-2}. Filtered water is used to prepare the ice flakes and blended with ice-cold filtered water.

The steam point is the boiling temperature of filtered water at a pressure of 1.0132×10^5 N.m^{-2}. Temperature interval between the steam point and the ice point is 100°C.

Kelvin (K), thermodynamic or absolute temperature scale on the name of Lord Kelvin is a scale, which does not depend on properties of any specific material. Lord Kelvin divided the temperature interval between the ice point and steam point into one-hundred divisions thus 1 K is equal to same interval of temperature as 1°C. The Kelvin unit is defined as the fraction (1/273.16) of the temperature of the water triple point. This definition was adopted by 13th CGPM in 1967.

ii.   Rankine and Fahrenheit scales

The Rankine and Fahrenheit scales are presently used in the United States and Britain. Steam tables, engineering data, etc. have been published utilizing the Rankine and Fahrenheit units. The steam point (212°F) and the ice point (32°F) are used on the Fahrenheit scale.

The Rankine cycle is a thermodynamic temperature corresponding to the Fahrenheit scale. The ice point is at 491.67°R while zero Fahrenheit is 459.67°R on the Rankine scale.

8.4.3 Temperature Measuring Instruments

These instruments are classified into two wide categories:

1. **Non-electrical methods**
   a. By using the change in a liquid volume with change of temperature
   b. By using the change of gas pressure with temperature change
   c. By using the pressure change of vapor with temperature change

2. **Electrical methods**
   a. Thermocouple
   b. Resistance thermometers
   c. By comparing the filament colors with that of the object whose temperature is to be found out, that is, optical pyrometers.
   d. By achieving the power received by radiation, that is, radiation pyrometers.

The following instruments are used for temperature measurements in industries:

1. Thermistors, 2. thermocouples, 3. RTD (Resistance Temperature Detectors), 4. IC sensors, 5. bimetallic indicators, 6. optical sensors such as: (a) pyrometers, (b) infrared detectors, (c) liquid crystals, 7. liquid bulb thermometers, 8. gas bulb thermometers.

The most popular instruments are the thermistors, the thermocouples, the RTD's and the IC sensors.

8.4.3.1 Classification of Thermometers

Thermometers are also classified as the following:

a. Expansion thermometers
   i. Liquid-in-glass thermometers
   ii. Bimetallic thermometers

b. Pressure thermometers
   i. Vapor pressure thermometers
   ii. Liquid-filled thermometers
   iii. Gas-filled thermometers

a. Thermocouple thermometers

b. Resistance thermometers

c. Radiation pyrometers

d. Optical pyrometers

8.4.4 High-temperature Thermometers

Mercury boils at 357°C at atmospheric pressure. In order to increase the range of mercury-in-glass thermometer for high temperature use, thermometer bore is expanded into a bulb having a space of about 20 times as that of the stem bore. The bore above the mercury with the bulb is filled with carbon dioxide or nitrogen at high pressure to prevent the boiling of mercury at high temperatures.

8.4.4.1 Use of Liquids other than mercury

In industries where the mercury leakage from the cracked bulb might cause large damage to the products, other liquids are utilized to fill in the thermometer. Other liquids are also used where the temperature range of the mercury-in-glass thermometer is not suitable. Table 8.1 shows liquid with their suitable temperature range.

Toluene, pentane, and ethyl alcohol can be used for liquid-in-glass thermometers. These liquids are colorless, tincture is added for easy reading. Due to low freezing point, these are suitable for low-temperature thermometers.

TABLE 8.1 Thermometer liquids with temperature range

Liquid	Temperature range (°C)
Mercury	−35 to +510
Alcohol	−80 to +70
Toluene	−80 to +100
Pentane	−200 to +30
Creosote	−5 to +200

8.4.5 Thermocouples Material

Thermocouple materials are divided into two groups depending on material cost, low-cost metal thermocouples and base metal thermocouples.

Table 8.2 shows thermocouples along with their specifications as per British standard BS4937. The schematic diagram of thermocouple is shown in Figure 8.2.

TABLE 8.2 Thermocouples as per British Standards

Types	Conductors (positive conductor first)	Manufactured to BS-4937 Part No.	Temperature tolerance class 2 thermocouple BS 5437	Service temperature
B	Platinum 30% Rhodium/ platinum 6% Rhodium	Part 7: 1974 (1981)	600 to 1700°C ±3°C	0–1500°C (1700°C) Better life expectancy at high temperature than types R&S
E	Nickel: chromium/ Constantan	Part 6: 1974 (1981)	−40 to +333 ±3°C 333 to 900°C ±0.75%	−200 to +850°C (1100°C) Resistant to oxidizing atmospheres
J	Iron/constantan	Part 3: 1973 (1981)	−40 to +333 2.5°C 150°C ±0.75%	−280 to +850°C (1100°C). Low cost suitable for general use
K	Nickel: chromium/ Nickel: aluminium (chrome/alumel) (C/A) (T_1/T_2)	Part 8: 1973 (1981)	+333°C±2.5°C 333 to 1200°C +0.75%	−200 to +1100°C (1300°C). Good general purpose. Best in oxidizing atmosphere
N	Nickel: Chromium: Silicon/nickel: Silicon: magnesium	Part 8: 1986	−40 to +333°C±2.5°C 333 to 1200°C ±0.75%	0–1100°C to (−270°C to +1300°C) Alternative to type K
R	Platinum 13% Rhodium/ platinum	Part 2: 1973 (1981)	0 to 600°C ±1.5°C 600 to 1600°C ±0.25%	0–1500°C (1650°C) High temperature Corrosion resistant
S	Platinum: 10% rhodium/platinum	Part 1: 1973 (1981)	0 to 600°C ±1.5°C 600 to 1600°C ±0.25%	Type R is more stable than type S
T	Copper/ Constantan (Copper/advance) (Cu/Con)	Part 5: 1974 (1981)	−40 to +375°C 1°C	−250 to 400°C (500°C). High resistance to corrosion by water

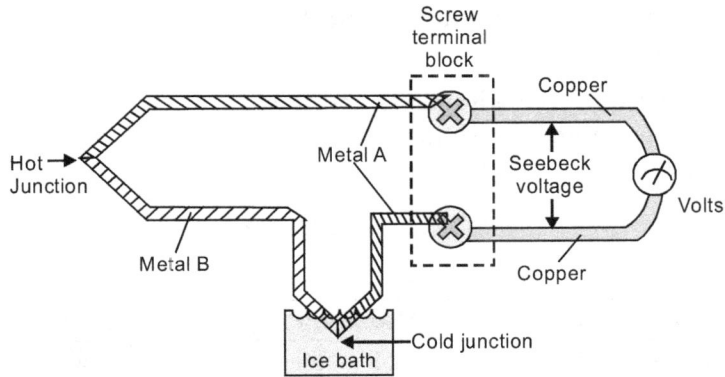

FIGURE 8.2 Thermocouple diagram

The recommended colors of thermocouple wire are shown in Table 8.3.

TABLE 8.3 Recommended colors of thermocouple wire

8.4.6 Temperature Calibration

There are two types of baths for calibration depending upon the temperature limit:

i.   water bath calibration

ii. sand bath calibration.

8.4.7 Comparison of Advantages and Disadvantages of Measurement Systems

Table 8.4 gives the advantages and disadvantages of three types of temperature measurement systems. The final selection of the technology depends upon design time, cost, circuit size, etc.

TABLE 8.4

a. Advantages

Thermocouple	Thermistor	RTD
Self-powered	High output	Most stable
Simple	Fast	Most accurate
Inexpensive	Tow-wire ohm measurement	Most linear than thermocouple
Variety of physical forms		

b. Disadvantages

Thermocouple	Thermistor	RTD
Non-linear for a wide range	Non-linear	Expensive
Low voltage	Limited temperature range	Slow
Reference required	Fragile	Current source required
Least stable	Current source required	Small resistance change
Least sensitive	Self-heating	

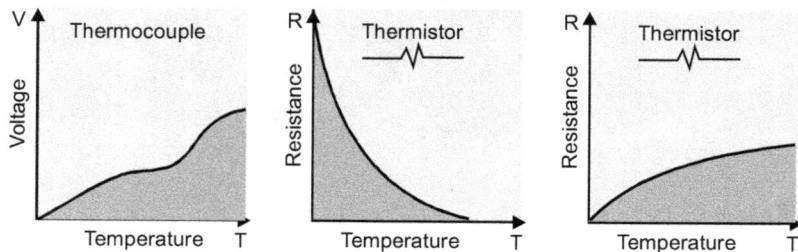

FIGURE 8.3 Advantages and disadvantages of three systems of temperature measurement with their differences

EXAMPLE 8.2

The changes of density of mercury are linear with the temperature as, $\rho_m = 14277.5 - 2.5\ T$ (K). Due to the influence of temperature, the same pressure difference will be measured by different manometer heights. Suppose in New York City, on a hot summer day the temperature is 40°C and the pressure is the same as the pressure measured on a cold winter day of –10°C. What will be the present deviation in the manometer reading?

Solution: Pressure is same for both cases,

$$\therefore\quad \rho_1 g h_1 = \rho_2 g h_2,$$

and

$$\frac{h_2}{h_1} = \frac{\rho_1}{\rho_2}$$

The present deviation in height may be expressed as follows:

$$Err\% = \frac{h_1 - h_2}{h_1} \times 100$$

$$Err\% = \left(1 - \frac{\rho_1}{\rho_2}\right) \times 100$$

As $T_1 = 313$ K and $T_2 = 263$ K from given relation, $\rho_1/\rho_2 = 0.9908$. The percentage error becomes, Err% = 0.91%, which is less than 1 percentage.

8.5   PRESSURE MEASUREMENT

8.5.1 Definition of Pressure

The fluid exerts a force on the boundary of the system. The pressure is the force per unit area. There are three types of pressure measurements.

1. **Absolute pressure** is defined as the difference between the absolute zero pressure and the pressure at a specific point in a fluid. This can be measured by a barometer.

2. **Gauge pressure:** Pressure gauge is used to measure the difference between the local atmospheric pressure and unknown pressure.

3. **Differential pressure:** A pressure gauge is used to measure the difference between two unknown pressures.

8.5.2 Pressure Units

Pascal	Pa
Bar	bar
Standard atmosphere	atm
Kilogram force per square cm	Kgf/cm^2
Pound force per square cm	Inch Ib/in^2
Millimeter of water	mm H$_2$O
Millimeter of mercury	mm Hg
Inch of water	In H$_2$O
Inch of mercury	In Hg

8.5.3 Pressure Thermometers

Vapor, liquid, and gas pressure thermometers can be used. This fluid is heated and fluid pressure increases. The temperature can be measured by Bourdon pressure gauge. The thermometer bulb full of fluid is put in the area where the temperature needs to be measured.

Pressure thermometers are explained below:

i.   Vapor pressure thermometer

A vapor pressure thermometer is shown in Figure 8.4. The fluid bulb is fixed in the area where the temperature needs to be measured. Some of the fluid vaporizes and raises the vapor pressure, which is measured on the Bourdon tube. The relationship between vapor pressure and temperature of a volatile liquid is of exponential shape.

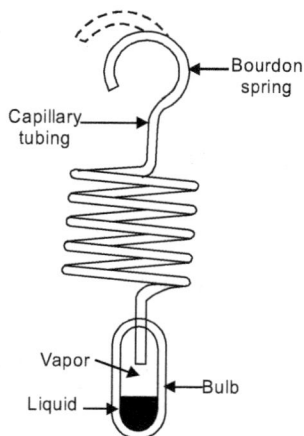

FIGURE 8.4 Vapor pressure thermometer

ii. Liquid-filled thermometer

A liquid-filled thermometer is shown in Figure 8.5. The liquid expansion causes the indicator to move on the dial. Alcohol, toluene, glycerin, and mercury have been successfully used for pressure changes between 3 and 100 bar. These thermometers are used for temperatures up to 650°C.

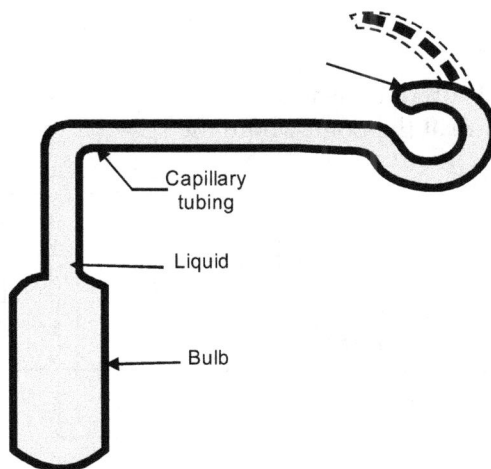

Capillary tubing

Liquid

Bulb

FIGURE 8.5 Liquid-filled thermometer

iii. Gas-filled thermometer

The gas-filled thermometer has almost the same range as the liquid-filled thermometer. Helium (He) and nitrogen (N_2) are insert gases, with low specific heat and high expansion coefficient. Therefore, He and N_2 gases are commonly used. The construction is same as the liquid-filled thermometer but with bigger bulbs.

These thermometers are commonly utilized for pressures less than 35 bar.

8.6 MEASUREMENT OF VELOCITY

The flow rate and velocity of flow are commonly needed for the control of the industrial processes and thermodynamic analysis of systems. The flow speed is generally measured indirectly by the following instruments:

i. pitot tube

ii. vane anemometer

iii. hot wire anemometer

iv. laser Doppler anemometer.

8.6.1 Pitot Tube

The velocity measurement of fluid is very commonly carried out by a pitot tube.

Principle

A pitot tube is used to measure the change between static pressure and stagnation pressure at the point where the speed is to be calculated (Figure 8.6).

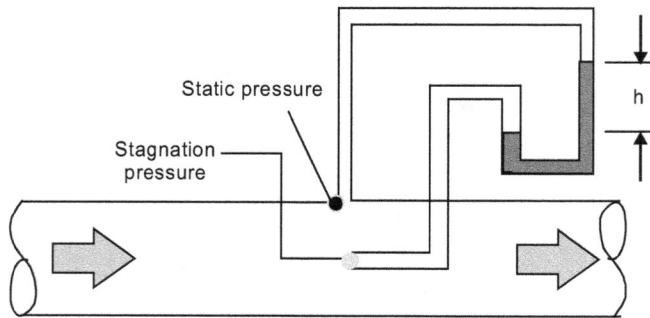

FIGURE 8.6 Pitot tube measures change of stagnation pressure

Figure 8.7 shows two types of tubes: a Brabbee tube and a Prandtl tube depending upon the tip shape. Fluid velocity near the tape is given as follows:

$$V = C\sqrt{2gh\left(\frac{\rho_m}{\rho} - 1\right)} \tag{8.1}$$

where C is the coefficient of speed and has to be calculated by calibrating the pitot tube. h = height of fluid in monometer. The velocity (V_1) of a perfect gas at the tip point,

$$V_1 = C\sqrt{\frac{2kRT_1}{K-1}\left(\left(\frac{P_0}{P_1}\right) - 1\right)} \tag{8.2}$$

where P_0 is the stagnation pressure, and (T_1, p_1) are the temperature and the pressure at a point where the speed is to be calculated.

The simplest pitot tube is a tube with an effective opening offer 3.125–6.35 mm diameter facing the approaching fluid. A normal upstream tab can be utilized for calculating the line pressure.

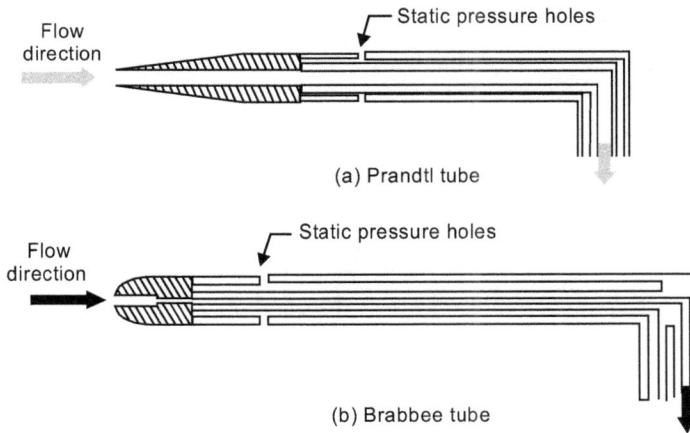

FIGURE 8.7 Pitot tube for velocity measurement

An industrial type of pitot tube cylindrical probe installed inside the air stream. At the upstream face, the fluid flow speed of the probe is taken to be zero. Velocity head transforms into short pressure, which is sensed by a small hole in upstream face. On the other side, a small hole of the probe measure static pressure. The instrument measures the differential pressure.

Figure 8.8 shows a pitot tube with the tabs of measuring static pressure.

The pitot tube produces no loss of pressure in the flow stream. It is installed over a nipple in the pipe. If installed over an isolation value, it can be moved backward and forward over the stream to create the flow velocity profile. For high-speed flow streams, it is necessary to use tubes of high strength and stiffness.

A tube installed in a high-speed stream has the possibility to fluctuate and get broken. Pitot tubes are commonly used for low to medium velocity flows.

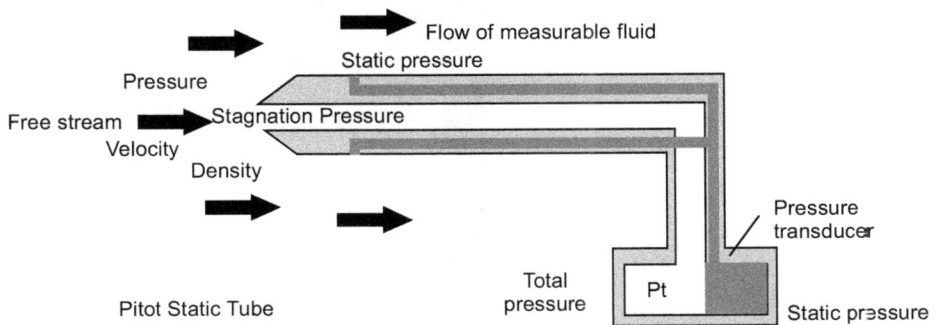

FIGURE 8.8 Pitot tube installation

EXAMPLE 8.3

The air velocity in a duct is measured as 40.1 m/s by a pitot tube and the recorded pressure is 0.12 m of water column. Take air density as 1.2 kg/m³, and determine the coefficient of velocity, C.

Solution: By applying Equation (8.1),

$$C = \frac{V}{\sqrt{2gh\left(\frac{\rho_m}{\rho} - 1\right)}}$$

$$C = \frac{40.1}{\sqrt{2 \times 9.81 \times 0.12 \left(\frac{1000}{1.2} - 1\right)}}$$

$$C = 0.905 \quad \textbf{Ans.}$$

8.6.2 Vane Anemometer

It is shown in Figure 8.9, the moving stream forces the vane to rotate. The strength of force increases with the velocity of air. The rotational velocity of the vane anemometer also increases. The vane anemometers are used to measure the speed of air in large flow areas. The anemometer calibration is carried out by measuring the rotational speed of vane at known wind speeds.

FIGURE 8.9 Vane anemometer

8.6.3 Hot Wire Anemometer

A thin platinum wire of 0.005 mm diameter and 1 mm length is electrically heated and put perpendicular into a flow area as shown in Figure 8.10. Heat is transferred from the hot wire anemometry to the fluid stream by conviction.

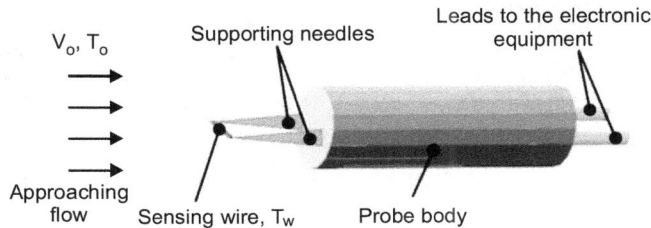

FIGURE 8.10 Probe of a hot wire anemometer

If the wire current and its temperature T_w are kept constant, the equation between the power input (to the wire) and the difference of temperature between the fluid and the wires is as follows:

$$\frac{I^2}{T_w - T_0} = A + B\sqrt{\rho V} \tag{8.3}$$

where R is the wire resistance, I is the current in the wire, ρ is the density of the free stream and V is the velocity of the free stream. B and A are constants and calculated by probe calibration curve. The hot wire anemometer is very suitable for measuring speeds in turbulent flow. The hot wire anemometer may record changes of velocity with a frequency (up-to-10^5 Hz), and speeds as small as 0.02 m/s. The hot wire anemometer is very suitable for small speed flows. It is used in the analysis of the boundary layer.

The hot wire anemometer has been used on a large scale for many years as a tool of research in fluid mechanics.

The hot wire anemometer is shown in Figures 8.11 and 8.12 as a small, electrically heated element. These elements are sensitive to heat transfer between the element and its surrounding and also structures and temperature changes can be sensed.

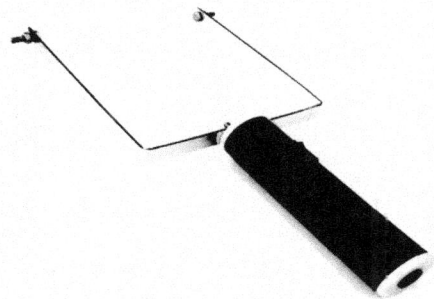

FIGURE 8.11 One type of hot wire anemometer

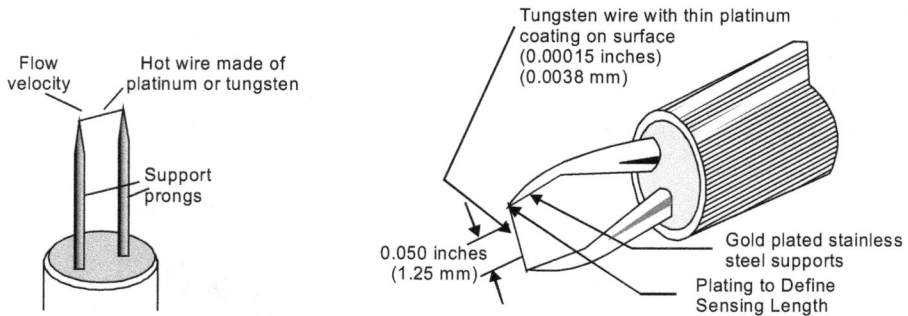

FIGURE 8.12 Another type of hot wire anemometer

EXAMPLE 8.4

The velocity data of air ($\rho = 1.2$ kg/m^3) flowing over a 24 cm, diameter pipe is recorded by the hot wire anemometer and given in the table below. Calculate the mean velocity, the mass flow rate of air, and the ratio of maximum to mean spread.

r (cm)	V (m/s)	r (cm)	V (m/s)
0	9.7	7	6.8
1	9.6	8	5.9
2	9.4	9	4.8
3	9.2	10	3.5
4	8.7	10.5	2.9
5	8.2	11	2.4
6	7.5	11.5	1.0

Solution: The mean velocity,

$$V_m = \frac{1}{R^2}\left[V_0 r_1^2 + \sum_{i=1}(V_i + V_{i+1})r_i\delta r_i\right]$$

$$V_m = \frac{1}{12^2}[9.7\times1^2 + 9.5\times2\times1\times1 + 9.3\times2\times2\times1 + 8.95\times2\times3\times1 + 8.45\times2$$

$$\times4\times1 + 7.85\times2\times5\times1 + 7.15\times2\times6\times1 + 6.35\times2\times7\times1 + 5.35\times2\times8\times1$$

$$+ 4.15\times2\times9\times1 + 3.2\times2\times10\times0.5 + 2.65\times2\times10.5\times0.5 + 1.7\times2\times11\times0.5]$$

At each strip, the velocity is the rate of the top and the bottom speeds. Thus the mean speed is $V_m = 4.71$ m/s, $V_{max} = 9.7$ m/s.

8.6.3.1 Principle of Operation

Consider a thin wire fixed to probes and subjected to a velocity U as shown in Figure 8.13. As current is passed through the wire, heat is produced (I^2R_W). This is balanced with loss of heat to the surrounding under equilibrium condition. With change of velocity, the coefficient of convective heat transfer will change, the temperature of the wire will alter and finally arrive at a new equilibrium state.

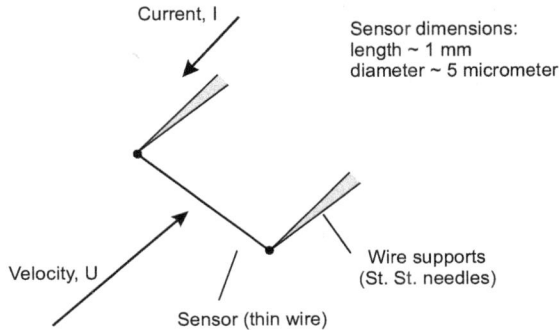

Current, I

Sensor dimensions:
length ~ 1 mm
diameter ~ 5 micrometer

Velocity, U

Wire supports
(St. St. needles)

Sensor (thin wire)

FIGURE 8.13 Principle of operation of hot wire anemometer

8.6.3.2 Classification of Hot Wire Anemometer

Based on the number of sensors, the classification of anemometer is shown in Figure 8.14.

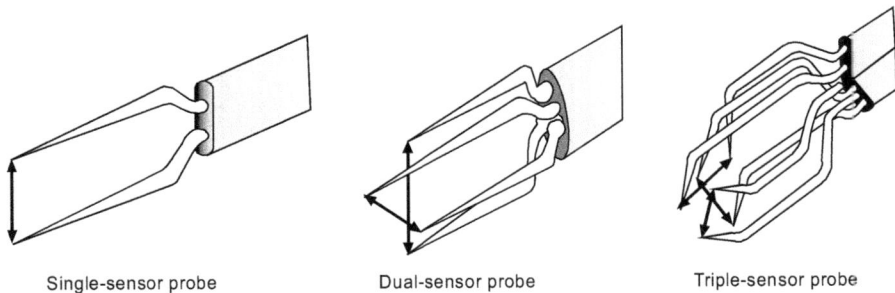

Single-sensor probe   Dual-sensor probe   Triple-sensor probe

FIGURE 8.14 Classification of hot wire probes

8.6.4 Laser Doppler Anemometer (LDA)

In the case of a hot wire anemometer, dirt inside the flow may be deposited on the wire acting as insulation, or because of high wire temperature, fluid may be spoiled in the measurement area. Furthermore, because of the probe size, the floor may be distributed under such objectives. LDA can be used for perfect measurement of speed and turbulence in liquid and gas flows.

As shown in Figure 8.15, a fixed at wavelength laser acts as a light source and the visual components divided the beam of the layer into a second beam and a reference beam and then these beams intersect at measurement particle.

The lighter frequency is dispersed by moving particle by a quantity proportionate to the practical velocity. Laser Doppler anemometer uses this principle in measuring the particle speed. Each of the frequency following filter closes onto the modification frequency to obtain the frequency of Doppler. The frequency of Doppler is linear regarding the speed over the geometry of the visual system.

The particles should be of small size sufficient to pursue the flow. The inflow of liquid-like particles is normally attended, but the inflow of gas, particle with size 10^{-6} m, are grown. To transmit the light beam, the fluid surroundings should be diaphanous. The applications of a laser Doppler anemometer include flow inside engine cylinders, pipe flows, combustion process, and flow between blades of the pump impeller.

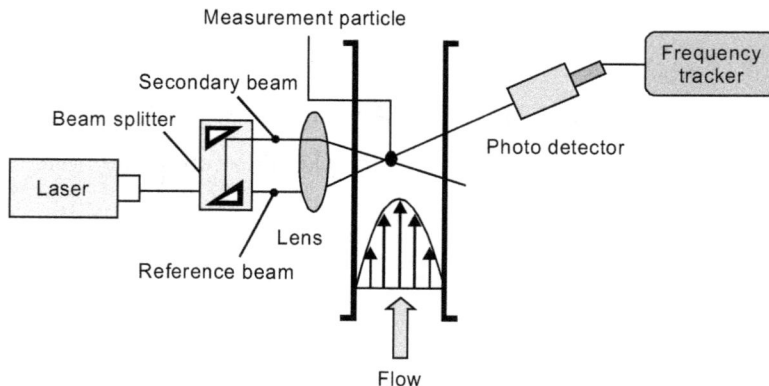

FIGURE 8.15 Operation of laser Doppler anemometer

8.7   FLOW MEASUREMENT

For a gaseous or liquid matter, flow can be measured as either the volume flow rate or the mass flow rate, per unit time. A mass flow rate gives more precise measurement.

8.7.1 Mass Flow Rate

The mass flow rate is measured by different techniques depending upon the fluid state, that is, gaseous, solid, or liquid.

8.7.1.1 Variable Area Flow Meter (Rotameter)

The area of orifice is used in rotameters to measure the flow rate. It is inexpensive, reliable, and used widely in industry, accounting for around 20% of all flow meters sold. For its use in automatic control charts, rotameter is combined with a row of fibers, which gives the float location by sensing the light emitted from it.

Figure 8.16 shows this instrument, which comprises a glass tube with tapered shape and a float is made of stainless steel that moves freely in side of the tube. The fluid flows over the float, the forces establish a balance float location in the inside of the tube.

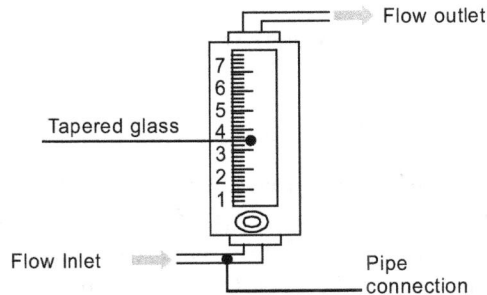

FIGURE 8.16 Schematic of rotameter

The device should be fixed vertically, and basically, the motion of float is linear with the flow rate. Variable area flow meters are not suitable for high pressures and liquids with large particles. The inaccuracy of the low-cost device is usually ±5%; however, more costly rotameter offers precision as high as ±0.5%.

A simple rotameter (Figure 8.17) consists of a tapered glass tube and a float that takes a steady position when flooded. The float location is a measure of the flow rate.

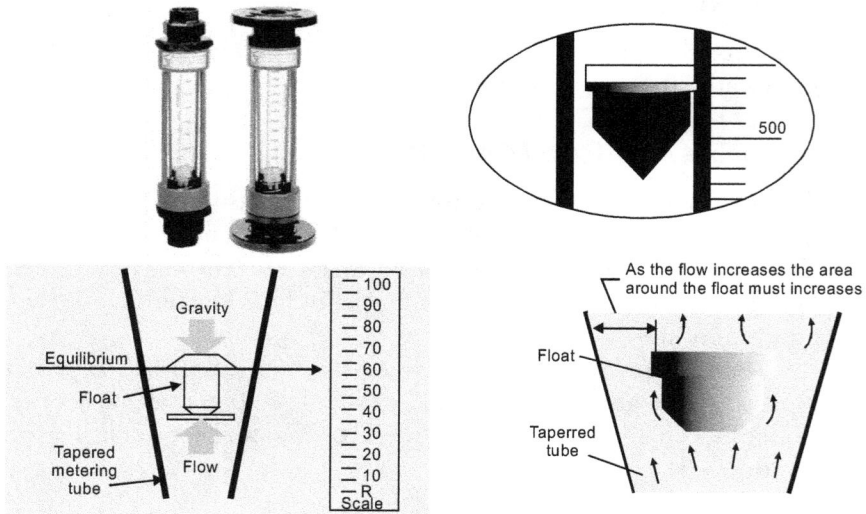

FIGURE 8.17 Flow meters (Rotameters)

EXAMPLE 8.5

At standard conditions (p = 1 atm, T = 20°C), a Variable Area Flow meter is calibrated for water, (ρ = 998 kg/m³). However, the rotameter is used to measure the oil flow rate (ρ = 880 kg/m³) and the scale on the rotameter reads (12 L/s). Calculate the oil flow rate.

Solution: The equation of flow rate in the rotameter is $\dot{V} = A \times V$, where A is the flow area between the float and the tube with a tapered shape. Thus, the ratio flow rate becomes $\dot{V}_0 / \dot{V}_w = V_0 / V_w$. The float of aerodynamic suspension is got by the same value of pressure drop through the float, $\Delta p_w = \Delta p_0$. Assuming that the pressure drop is commensurate with V_2,

$$\rho_w V_w^2 = \rho_0 V_0^2$$

$$\dot{V}_0 / \dot{V}_w = \sqrt{\frac{\rho_w}{\rho_0}}$$

Hence,

$$\dot{V}_0 = 12 \times 1.066$$

Or

$$\dot{V}_0 = 12.79 \text{ L/s. } \textbf{Ans.}$$

8.7.1.2 Orifice Plate

The orifice plate is a metal disc having a hole in its center, which is installed inside the pipe, carrying the flowing fluid. Orifice plates are cheap and available in a board range of sizes. As a result, they represent around 50% of the devices used in the industry for measuring the volumetric flow rate. One restriction of the orifice plate is that its inaccuracy is usually ±2% and may approach ±5%. Furthermore, pressure loss caused in the measured fluid flow is between 50 and 90% of the pressure change, $(P_1 - P_2)$. The other orifice plate problems are a progressive alteration in the coefficient of discharge through the hole edges and a tendency for particles inside the flowing fluid to block the hole. There may be decrease in diameter progressively as the particles stick. The last problem can be decreased by using an orifice plate with a hole at the center. If this hole is near to the pipe bottom, solids inside the flowing fluid incline to be swept across, and particles build up is decreased.

The same problem appears when there are gas or vapor bubbles in the flowing liquid. This difficulty can be removed by installing the orifice plate in the vertical portion of pipeline.

The hole shapes in orifice plates may be eccentric, concentric or segmental as illustrated in Figure 8.18. Orifice plates are damaged by erosion.

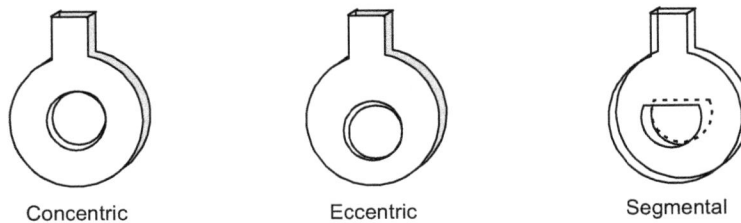

Concentric   Eccentric   Segmental

FIGURE 8.18 Eccentric, concentric, and segmental orifice plates

$$Q = S_0 C_d \sqrt{\frac{2(P_1 - P_2) + pg(z_1 - z_2)}{\rho[1 - (d_0/d_1)^4]}} \tag{8.4}$$

or,

$$Q = S_0 C_d \sqrt{\frac{2g(\Delta h) + (z_1 - z_2)}{[1 - (d_0/d_1)^4]}} \tag{8.5}$$

The discharge coefficient to C_d for a particular orifice meter is a function of the portion of the pressure tapes, the diameter ratio of the orifice to the inner pipe diameter, the Reynolds number (R_e) in the pipeline, and the orifice plate thickness.

As shown in Figure 8.19, an orifice meter consists of a thin plate with an aperture in its center. The orifice aperture is commonly circular. The abrupt constriction of area in these devices produces a high loss of pressure compared with the other two devices. The downstream pressure is measured at a place of $0.5\,D_1$ length and upstream pressure is measured at a place D_1 length.

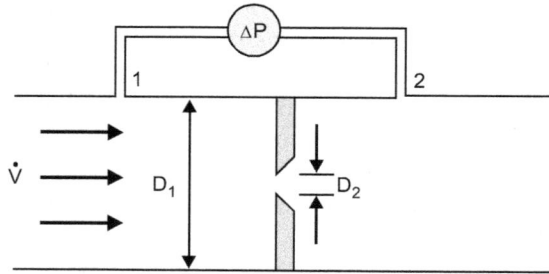

FIGURE 8.19 Orifice meter

As a function of pressure drop, the measured flow rate over the orifice plate is expressed as follows:

$$\dot{V} = K \times A_2 \sqrt{2\Delta P/\rho} \qquad (8.6)$$

where K is the coefficient of the orifice plate and, depends upon the plate diameter ratio and the Reynolds number (R_e) of the flow that is expressed as, $R_e = \rho V D_1/\mu$.

EXAMPLE 8.6

A pipe of diameter (30 cm) loads oil, (ρ = 880 kg/m^3, μ = 0.799 kg/m^2), and to determine the flow rate an orifice (diameter = 15 cm) is fitted to the pipe. A mercury manometer is used to read the pressure drop through the office as 0.95 m. Calculate the flow rate of oil.

Solution:

$$\frac{D_2}{D_1} = 0.5, \text{ and assume high Reynolds number flow.}$$

From Figure 8.20,

$$K = 0.62$$

$$\Delta p = \rho mgh = 13.6 \times 9.81 \times 0.95 = 126.745 \text{ kPA.}$$

$$A_2 = 3.14 \times \frac{0.15^2}{4} = 0.0176 \text{ m}^2.$$

These values are substituted into Equation 8.6

$$\dot{V} = 0.62 \times 0.0176 \times \sqrt{\frac{2 \times 126745}{880}} = 0.185 \text{ m}^3/\text{s}$$

The condition of R_e must be checked.

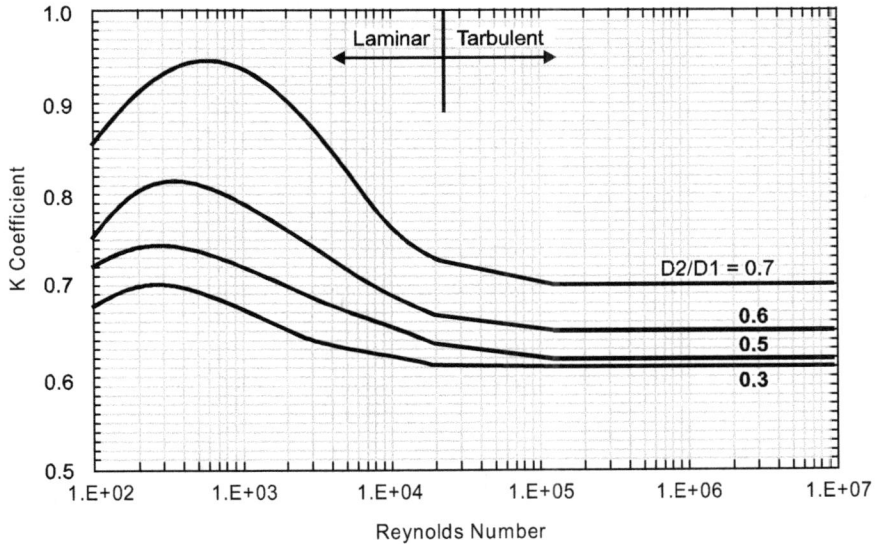

FIGURE 8.20 Coefficients of discharge of the orifice [J. H. Perry, 1984]

$$A_1 = 0.07 \text{ m}^2, \ V = \frac{0.185}{0.07} = 2.645 \text{ m/s},$$

And

$$R_e = \frac{880 \times 2.645 \times 0.3}{0.799} = 874.$$

With respect to Figure 8.20,

$K = 0.72$, and the corrected flow rate becomes,
$$\dot{V} = 0.214 \text{ m}^3/\text{s}.$$

8.7.1.3 Venturi Meter

The Venturi is an accurately machined tube. This shows measurement inaccuracy of just ($\pm 1\%$). However, it requires machining, which means cost. Pressure loss in the measuring system is (10–15%) of the pressure difference ($P_1 - P_2$).

The first use of Venturi meter was by J.B Venturi in 1797. As shown in Figure 8.21, reducing the flow cross-sectional area, some pressure head is transformed into speed.

The head differential has to be measured between the throat part and the upstream to evaluate the flow rate.

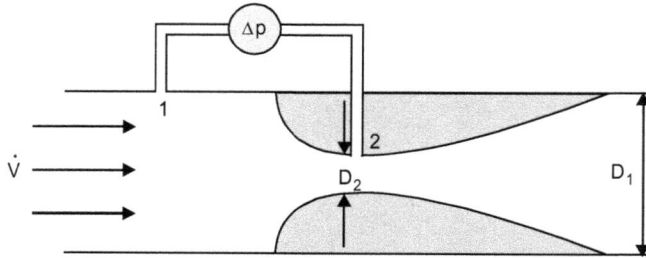

FIGURE 8.21 Cross-section of a Venturi meter

$$\dot{V} = CA_2 = \sqrt{\frac{2\Delta P}{\rho(1-\beta^4)}} \tag{8.7}$$

where P is the density of the fluid, and $\left(\beta = \dfrac{D_2}{D_1}\right)$, for Venturi meter, $0.25 < \beta < 0.50$.

C is the coefficient of discharge and changes in the range of 0.935–0.988.

The upstream part closeness angle is $21°$ with the axis of pipe and the diverging part has an angle $5°-7°$. In linking the Venturi meter to a pipeline, the upstream part of the Venturi should have a distance $10D_1$ straight pipe. The total losses in a Venturi are approximately between 10 and 20 percentage of the total pressure drop.

EXAMPLE 8.7

A venturi meter was designed for measuring the flow rate of water in horizontal pipes of diameter 300 cm. The evaluated discharge over the pipe is 15 m³/s, and the pressure drop is restricted to 250 kPa at the venturi. Determine the diameter of the venturi throat. Assume that the coefficient of discharge is 0.95.

Solution: From Equation 8.7, the relationship between A_2 and β is,

$$\frac{\sqrt{1-\beta^4}}{A_2} = \frac{0.95\sqrt{\dfrac{2\times 25000}{1000}}}{15} = 1.416$$

Substituting,

$$D_2 = 0.948(1 - \beta^4)^{0.25}$$

This relation is determined for A_2 by method of trial and error. Suppose $\beta = 0.35$, calculate (D_2), $D_2 = 0.944$ m, and $\beta = 0.314$, after that finally diameter of venturi throat is $D_2 = 0.945$ m.

8.7.2 Calibration of Flow Meters

The first consideration in selecting calibration method for flow measuring devices is to establish the precision level necessary. The calibration system should not be costly and should satisfy the statutory requirements needed.

The flow measurement precision is influenced greatly by the conditions of flow and the flowing fluid characteristics. Flow measuring devices are calibrated on-sight in the field. The calibration is achieved under the conditions of practical flow, which are difficult to re-produce in a laboratory. For the validity of calibration, it is necessary to replicate flow calibration tests so that the same value of reading is obtained in two sequential tests.

If onsite calibration is not possible or is not of sufficient precision, the device may be sent for calibration utilizing equipment supplier or device manufacturers or other calibration organization. However, this can be costly. Moreover, the calibration facility does not repeat the normal working conditions of the meter examined, and suitable compensation for variance between conditions of calibration and conditions of natural utilization should be applied.

8.8   MEASUREMENT SYSTEM ERROR

In a perfect measurement system, the measured value should be the same as the correct value. The system precision can be defined as nearness of the measured value to the correct value. A completely precision system is a theoretical system. Processor of a real system is quantified utilizing the error of measurement system (E),

where $E = $ [measured value] $-$ [true value].

$E = $ [system output] $-$ [system input].

REVIEW QUESTIONS

1. What is a measurement system? How will you define its performance? Explain the following terms:

    i. Calibration

    ii. Measurement uncertainty

    iii. Sensitivity and repeatability

    iv. System error

2. Explain the following:

    i. Temperature scales

    ii. Temperature measuring instruments

    iii. Thermocouples

3. How will you select your temperature measurement instrument?

4. Write a detailed note on pressure thermometer.

5. Explain the following devices for velocity measurement:

    a. Pitot tube

    b. Vane anemometer

    c. Hot wire anemometer

    d. Laser Doppler anemometer

6. Compare the following flow meters:

    a. Rotameter

    b. Orifice plate

    c. Venturi meter

NUMERICAL EXERCISES

1. During temperature measurement, it is found that a thermometer gives the same temperature in °C and °F. Express the temperature value in R.

    (419.67 R)

2. A platinum resistance thermometer has resistance of 2.8 ohms at 0°C and 3.8 ohms at 100°C. Calculate the temperature when the resistance is indicated as 5.8 ohm. **(300°C)**

3. A certain thermometer using pressure as the thermometric property gives the value of pressure as 1.86 and 6.81 at ice point and steam point, respectively. The temperature of ice point and steam point are assigned the number 32 and 212, respectively.

   Determine the temperature corresponding to p −2.5 if the temperature t is defined in terms of pressure p as $t = a \ln p + b$, where a and b are constants. **(73°C)**

4. A turbine is supplied with steam at a pressure of 20 bar gauge. After expansion in the turbine, the steam passes to condenser, which is maintained at a vacuum of 250 mm of mercury by means of pumps. Express the inlet and exhaust steam pressure in N/m^2 and kPa.
   (2101396.2 N/m^2 or 2101.4 kPa
   68024 N/m^2 or 68 kPa)

5. A vessel of cylindrical shape of 50 cm diameter and 75 cm height contains 4 kg of gas. The pressure gauge mounted on the vessel indicates 620 mm of Hg above atmosphere. If the borometer reading is 760 mm of Hg, calculate the absolute pressure of gas in bar. Also determine the density and specific volume of the gas. **(1.84 bar, 27.21 kg/m^3, 0.036 m^3/kg)**

Thermodynamic Tables SI Units

Table	
Table A-1	Atomic or Molecular Weights and Critical Properties of Selected Elements and Compounds
Table A-2	Properties of Saturated Water (Liquid-Vapor): Temperature Table
Table A-3	Properties of Saturated Water (Liquid-Vapor): Pressure Table
Table A-4	Properties of Superheated Water Vapor
Table A-5	Properties of Compressed Liquid Water
Table A-6	Properties of Saturated Water (Solid-Vapor): Temperature Table
Table A-7	Properties of Saturated Refrigerant 22 (Liquid-Vapor): Temperature Table
Table A-8	Properties of Saturated Refrigerant 22 (Liquid-Vapor): Pressure Table
Table A-9	Properties of Superheated Refrigerant 22 Vapor
Table A-10	Properties of Saturated Refrigerant 134a (Liquid-Vapor): Temperature Table
Table A-11	Properties of Saturated Refrigerant 134a (Liquid-Vapor): Pressure Table
Table A-12	Properties of Superheated Refrigerant 134a Vapor
Table A-13	Properties of Saturated Ammonia (Liquid-Vapor): Temperature Table
Table A-14	Properties of Saturated Ammonia (Liquid-Vapor): Pressure Table
Table A-15	Properties of Superheated Ammonia Vapor
Table A-16	Properties of Saturated Propane (Liquid-Vapor): Temperature Table
Table A-17	Properties of Saturated Propane (Liquid-Vapor): Pressure Table
Table A-18	Properties of Superheated Propane Vapor
Table A-19	Properties of Selected Solids and Liquids: c_p, ? and K
Table A-20	Ideal Gas Specific Heats of Some Common Gases
Table A-21	Variation of c_p with Temperature for Selected Ideal Gases

Table A-22	Ideal Gas Properties of Air
Table A-23	Ideal Gas Properties of Selected Gases
Table A-24	Thermochemical Properties of Selected Substances at 298 K and 1 atm
Table A-25	Standard Molar Chemical Exergy of Selected Substances at 298 K and $p0$

TABLE A-1 Atomic or Molecular Weights and Critical Properties of Selected Elements and Compounds

Substance	Chemical Formula	M (kg/kmol)	T_c (K)	P_c (bar)	$Z_c = \dfrac{P_c v_c}{RT_c}$
Acetylene	C_2H_2	26.04	309	62.8	0.274
Air (equivalent)	–	28.97	133	37.7	0.284
Ammonia	NH_3	17.03	406	112.8	0.242
Argon	Ar	39.94	151	48.6	0.290
Benzene	C_5H_6	78.11	563	49.3	0.274
Butane	C_4H_{10}	58.12	425	38.0	0.274
Carbon	C	12.01	–	–	–
Carbon dioxide	CO_2	44.01	304	73.9	0.276
Carbon monoxide	CO	28.01	133	35.0	0.294
Copper	Cu	63.54	–	–	–
Ethane	C_2H_6	30.07	305	48.8	0.285
Ethyl alcohol	C_2H_5OH	46.07	516	63.8	0.249
Ethylene	C_2H_4	28.05	283	51.2	0.270
Helium	He	4.003	5.2	2.3	0.300
Hydrogen	H_2	2.016	33.2	13.0	0.304
Methane	CH_4	16.04	191	46.4	0.290
Methyl alcohol	CH_3OH	32.04	513	79.5	0.220
Nitrogen	N_2	28.01	126	33.9	0.291
Octane	C_8H_{18}	114.22	569	24.9	0.258
Oxygen	C_3H_8	32.00	154	50.5	0.290
Propane	O_2	44.09	370	42.7	0.276
Propylene	C_3H_6	42.08	365	46.2	0.276
Refrigerant 12	CCl_2F_2	120.92	385	412	0.278
Refrigerant 22	$CHClF_2$	86.48	369	49.8	0.267
Refrigerant 134a	CF_3CH_2F	102.03	374	40.7	0.260
Sulfur dioxide	SO_2	64.60	431	78.7	0.268
Water	H_2O	18.02	647.3	220.9	0.233

TABLE A-2 Properties of Saturated Water (Liquid-Vapor): Temperature Table

| Temp. °C | Press. Bar | Specific Volume m³/kg | | Internal Energy kJ/kg | | Enthalpy kJ/kg | | | Entropy kJ/kg K | | Temp. °C |
		Sat. Liquid $v_f \times 10^3$	Sat. Vapor v_g	Sat. Liquid u_f	Sat. Vapor u_g	Sat. Liquid h_f	Evap. h_{fg}	Sat. Vapor h_g	Sat. Liquid s_f	Sat. Vapor s_g	
0.01	0.00611	1.000	206.136	0.00	2375.3	0.01	2501.3	2501.4	0.0000	9.1562	0.01
4	0.00813	1.001	157.232	16.77	2350.9	16.78	2491.9	2508.7	0.0610	9.0514	4
5	0.00872	1.0001	147.120	20.97	2382.3	20.98	2489.6	2510.6	0.0761	9.0257	5
6	0.00935	1.0001	137.734	25.19	2383.6	25.20	2487.2	2512.4	0.0912	9.0003	6
8	0.01072	1.0001	120.917	33.59	2386.4	33.60	2482.2	2516.1	0.1212	8.9501	8
10	0.01228	1.0004	106.379	42.00	2389.2	42.01	2477.7	2519.5	0.1510	8.9008	10
11	0.01312	1.0004	99.857	40.20	2390.5	46.20	2475.4	2521.6	0.1658	8.8765	11
12	0.01402	1.0005	93.754	50.41	2391.9	50.41	2473.0	2523.4	0.1806	8.8524	12
13	0.01497	1.0007	88.124	54.60	2393.3	54.60	2470.7	2525.3	0.1953	8.8285	13
14	0.01598	1.0008	82.848	58.79	2394.7	58.80	2468.3	2527.1	0.2099	8.8045	14
15	0.01705	1.0009	77.926	62.99	2396.1	62.99	2465.9	2528.9	0.2245	8.7814	15
16	0.01818	1.0004	73.333	67.IS	2397.4	67.19	2463.6	2530.8	0.2390	8.7582	16
17	0.01938	1.0012	69.044	71.38	2398.8	71.38	2461.2	2532.6	0.2535	8.7351	17
18	0.02064	1.0014	65.038	75.57	2400.2	75.58	2458.8	2534.4	0.2679	8.7123	18
19	0.02198	1.0016	61.293	79.76	2401.6	79.77	2456.5	2536.2	0.2523	8.6897	19
20	0.02339	1.0018	57.791	83.95	2402.9	83.96	2454.1	2538.1	0.2966	8.6672	20
21	0.02487	1.0020	54.514	85.14	2404.3	88.14	2451.8	2539.9	0.3109	8.6450	21
22	0.02645	1.0022	51.447	92.32	2405.7	92.33	2449.4	2541.7	0.3251	8.6229	22
23	0.02810	1.0024	48.574	96.51	2407.0	9652	2447.0	2543.5	0.3393	8.6011	23
24	0.02985	1.0027	45.883	100.70	2405.4	100.70	2444.7	2545.4	0.3534	8.5794	24

TABLE A-2 (Continued)

Temp. °C	Press. Bar	Specific Volume m³/kg Sat. Liquid $v_f \times 10^3$	Specific Volume m³/kg Sat. Vapor v_s	Internal Energy kJ/kg Sat. Liquid u_f	Internal Energy kJ/kg Sat. Vapor u_g	Enthalpy kJ/kg Sat. Liquid h_f	Enthalpy kJ/kg Evap. h_{fg}	Enthalpy kJ/kg Sat. Vapor h_g	Entropy kJ/kg K Sat. Liquid s_f	Entropy kJ/kg K Sat. Vapor s_g	Temp. °C
25	0.03169	1.0029	43.360	104.88	2409.8	104.59	2442.3	2547.2	0.3674	8.5580	25
26	0.03363	1.0032	40.994	109.06	2411.1	109.07	2439.9	2549.0	0.3814	8.5367	26
27	0.03567	1.0035	38.774	113.25	2412.5	113.25	2437.6	2550.8	0.3954	8.5156	27
28	0.03782	1.0037	36.690	117.42	2413.9	117.43	2435.2	2552.6	0.4093	8.4946	28
29	0.04008	1.0040	34.733	121.60	2415.2	121.61	2432.5	2554.5	0.4231	8.4739	29
30	0.04246	1.0043	32.594	125.78	2416.6	125.79	2430.5	2556.3	0.4369	8.4533	30
31	0.04496	1.0046	31.165	129.96	2418.0	129.97	2428.1	2558.1	0.4507	8.4329	51
32	0.04759	1.0050	1.0050	134.14	2419.3	134.15	2425.7	2559.9	0.4644	8.4127	32
33	0.05034	1.0053	1.0053	138.32	2420.7	138.33	2423.4	2561.7	0.4781	8.3927	33
34	0.05324	1.0056	1.0056	142.50	2422.0	142.50	2421.0	2563.5	0.4917	8.3728	34
35	0.05628	1.0060	1.0060	146.67	2423.4	146.68	2418.6	2565.3	0.5053	8.3531	35
36	0.05947	1.0063	1.0063	150.85	2424.7	150.86	2416.2	2567.1	0.5188	8.3336	36
38	0.06632	1.0071	1.0071	159.20	2427.4	159.21	2411.5	2570.7	0.5458	8.2950	38
40	0.07354	1.0078	1.0078	167.56	2430.1	167.57	2406.7	2574.3	0.5725	8.2570	40
45	0.09593	1.0099	1.0099	188.44	2436.8	188.45	2394.8	2583.2	0.6387	8.1648	45
50	0.1335	1.0121	1.0121	209.32	2443.5	209.33	2382.7	2592.1	0.7038	8.0763	50
55	0.1576	1.0146	1.0146	230.21	2450.1	230.23	1370.7	2600.9	0.7679	7.9913	35
60	0.1994	1.0172	1.0172	111.11	2456.6	251.13	2358.5	2609.6	0.8312	7.9096	60
65	0.2503	1.0199	1.0199	272.02	2463.1	272.06	2346.2	2618.3	0.8935	7.8310	63
70											

TABLE A-2 (Continued)

| Temp. °C | Press. Bar | Specific Volume m³/kg | | Internal Energy kJ/kg | | Enthalpy kJ/kg | | | Entropy kJ/kg K | | Temp. °C |
		Sat. Liquid $v_f \times 10^3$	Sat. Vapor v_g	Sat. Liquid u_f	Sat. Vapor u_g	Sat. Liquid h_f	Evap. h_{fg}	Sat. Vapor h_g	Sat. Liquid s_f	Sat. Vapor s_g	
75	0.3119	1.0228	1.0228	292.95	2469.6	292.98	2333.8	2626.8	0.9549	7.7553	70
80	0.3858	1.0259	1.0259	313.90	2473.9	313.93	2321.4	2635.3	1.0155	7.6824	75
85	0.4139	1.0291	1.0291	334.86	2482.2	334.91	2308.8	2643.7	1.0753	7.6122	80
90	0.5783	1.0325	1.0325	355.84	2488.4	355.90	2296.0	2651.9	1.1343	7.5445	83
95	0.7014	1.0380	1.0380	376.85	2494.5	376.92	2283.2	2660.1	1.1923	7.4791	90
	0.8455	1.0397	1.0397	397.88	2500.6	397.96	2270.2	2668.1	1.2500	7.4159	95
100	1.014	1.0435	1.0435	418.94	2506.5	419.04	2.2570	2676.1	1.3069	7.3549	100
110	1.433	1.0516	1.0516	461.14	2518.1	461.30	2230.2	2691.5	1.4183	7.2387	110
120	1.985	1.0603	1.0603	503.50	2529.3	503.71	2.2026	2706.3	1.5276	7.1296	120
130	2.701	1.0697	1.0697	546.02	2539.9	546.31	2174.2	2720.5	1.6344	7.0269	130
140	3.613	1.0797	1.0797	588.74	2550.0	589.13	2144.7	2733.9	1.7391	6.9299	140
150	4.758	1.0905	1.0905	631.68	2559.5	632.20	2114.3	2746.5	1.8418	6.8379	150
160	6.178	1.1020	1.1020	674.86	2568.4	675.55	2082.6	2758.1	1.9427	6.7503	160
170	7.917	1.1143	1.1143	718.33	2576.5	719.21	3H95	2768.7	2.0419	6.6663	170
180	10.02	1.1274	1.1274	761.09	2583.7	763.23	2015.0	2778.2	2.1396	6.5857	180
190	12.54	1.1414	1.1414	806.19	25900	807.62	1978.8	2786.4	2.2359	6.5079	190
200	15.54	1.1565	0.1274	850.65	2595.3	852.45	1940.7	2793.2	2.3309	6.4323	200
210	19.06	1.1726	0.1044	895.53	2599.5	897.76	1900.7	2798.5	2.4248	6.3585	210
220	33.18	1.1900	0.08619	940.87	2602.4	943.62	1858.5	2801.1	2.5178	6.2861	220
230	27.95	1.2088	0.07158	9867.4	76003.9	990.12	1813.8	2804.0	2.6099	6.2146	230
240	33.44	1.2291	0.04976	1033.2	2601.0	1037.3	1766.5	2503.8	2.7015	6.1437	240
250	39.73	1.2512	0.05013	1080.4	2602.4	1085.4	1716.2	2801.5	2.7927	6.0730	150
260	46.88	1.2755	0.04221	1128.4	1599.0	1134.4	1662.5	2796.6	2.8838	6.0019	260

TABLE A-2 (Continued)

Temp. °C	Press Bar	Specific Volume m³/kg		Internal Energy kJ/kg		Enthalpy kJ/kg			Entropy kJ/kg K		Temp. °C
		Sat. Liquid $v_f \times 10^3$	Sat. Vapor v_g	Sat. Liquid u_f	Sat. Vapor u_g	Sat. Liquid h_f	Evap. h_{fg}	Sat. Vapor h_g	Sat. Liquid s_f	Sat. Vapor s_g	
270	54.99	1.3023	0.03564	1177.4	2593.7	1184.5	1716.2	2789.7	2.9751	5.9301	270
280	64.12	1.3821	0.03017	1227.5	2586.1	1236.0	1543.6	2779.6	3.0668	5.8571	280
290	74.36	1.3656	0.02557	1278.9	2576.0	1289.1	1477.1	2766.2	3.1594	5.7821	290
300	85.81	1.4036	0.02167	1331.0	2563.0	1344.0	1404.9	2749.0	3.2534	5.7045	300
320	112.7	1.4988	0.01549	1444.6	2525.5	1461.5	1238.6	2700.1	3.4480	5.5362	320
340	145.9	1.6379	0.01080	1570.3	2464.6	1581.2	1027.9	2621.0	3.6584	5.3357	340
360	186.5	1.8975	0.006945	1725.2	2351.5	1760.5	720.5	2481.0	3.9147	5.0526	360
374.14	220.9	3.155	0.003155	2029.6	2029.6	2099.3	0	2099.3	4.4298	4.4298	374.14

TABLE A-3 Properties of Saturated Water (Liquid-Vapor): Pressure Table

Temp. °C	Press. bar	Specific volume m³/kg Sat. Liquid $v_f \times 10^3$	Sat. Vapor v_g	Internal Energy kJ/kg Sat. Liquid u_f	Sat. Vapor u_g	Enthalpy kJ/kg Sat. Liquid h_f	Evap. h_{fg}	Sat. Vapor h_g	Entropy kJ/kg K Sat. Liquid s_f	Sat. Vapor s_g	Press. bar
0.04	28.96	1.0040	34.800	121.45	2115.2	121.46	2432.9	2554.4	0.4226	8.4746	0.04
0.06	36.16	1.0064	23.739	151.53	2425.0	151.53	2415.9	2567.4	0.5210	8.3304	0.06
0.08	41.51	1.0084	18.103	173.87	2431.2	173.88	2403.1	2577.0	0.5926	8.2287	0.08
0.10	45.81	1.0102	14.614	191.82	2437.9	191.83	2391.8	2584.7	0.6493	8.1502	0.10
0.20	60.06	1.0172	7.649	251.38	2456.7	251.40	2358.3	2609.7	0.8320	7.9085	0.20
0.30	69.10	1.0223	5.229	289.20	2468.4	289.23	2336.1	2625.3	0.9439	7.7686	0.30
0.40	75.87	1.0265	3.993	317.53	2477.0	317.58	2319.2	2636.8	1.0259	7.6700	0.40
0.50	81.33	1.0300	3.240	340.44	2483.9	340.49	2305.4	2645.9	1.0910	7.5939	0.50
0.60	85.94	1.0331	2.732	359.79	2489.6	359.86	2293.6	2653.5	1.1453	7.5320	0.60
0.70	89.95	1.0360	2.385	376.63	2494.5	376.70	2283.3	2660.0	1.1919	7.4797	0.70
0.80	93.50	1.0380	2.087	391.58	2498.8	391.66	2274.1	2665.8	1.2329	7.4346	0.80
0.90	96.71	1.0410	1.869	405.06	2502.6	405.15	2265.7	2670.9	1.2695	7.3949	0.90
1.00	99.63	1.0432	1.694	417.36	2506.1	417.46	2258.0	2675.5	1.3026	1.3026	1.00
1.50	111.4	1.0528	1.159	466.94	2519.7	467.11	2226.5	2693.6	1.4336	7.2233	1.50
2.00	120.2	1.0605	0.8857	504.49	2529.5	504.70	2201.9	2706.7	1.5301	7.1271	2.00
2.50	127.4	1.0672	0.7187	535.10	2537.2	535.37	2181.5	2716.9	1.6072	7.0527	2.50
3.00	133.6	1.0732	0.6058	561.15	2543.6	561.47	2163.8	2725.3	1.6718	6.9919	3.00
3.50	138.9	1.0786	0.5243	583.95	2546.9	584.33	2148.1	2732.4	1.7275	6.9405	3.50
4.00	143.6	1.0836	0.4625	604.31	2553.6	604.74	2133.8	2738.6	1.7766	6.8959	4.00
4.50	147.9	1.0882	0.4140	622.25	2557.6	621.25	2120.7	2743.9	1.8207	6.8565	4.50

TABLE A-3 (Continued)

Temp. °C	Press. bar	Specific volume m³/kg Sat. Liquid $v_r \times 10^3$	Sat. Vapor v_s	Internal Energy kJ/kg Sat. Liquid u_f	Sat. Vapor u_g	Enthalpy kJ/kg Sat. Liquid h_f	Evap. h_{fg}	Sat. Vapor h_g	Entropy kJ/kg K Sat. Liquid s_f	Sat. Vapor s_g	Press. bar
5.00	151.9	1.0926	0.3749	639.68	2561.2	640.23	2108.5	2748.7	1.8607	6.8212	5.00
6.00	158.9	1.1006	0.3157	669.90	2567.4	670.56	2086.3	2736.8	1.9312	6.7600	6.00
7.00	165.0	1.1080	0.2729	696.44	2572.5	697.22	2066.3	2763.5	1.9922	6.7080	7.00
8.00	110.4	1.1148	0.2404	720.22	2576.8	721.11	2048.0	2769.1	2.0462	6.6628	8.00
9.00	175.4	1.1212	0.2150	741.83	2580.5	742.83	2031.1	2773.9	2.0946	6.6226	9.00
10.0	179.9	1.1273	0.1944	761.68	2583.6	762.81	2015.3	2778.1	1.1387	6.5863	10.0
15.0	198.3	1.1539	0.1318	843.16	2594.5	844.84	1947.3	2792.2	1.3150	6.4448	15.0
20.0	212.4	1.1767	0.09963	906.44	2600.3	908.79	1890.7	2799.5	2.4474	6.3409	20.0
25.0	224.0	1.1973	0.07998	959.11	2603.1	962.11	1841.0	2803.1	2.5547	6.2575	25.0
30.0	233.9	1.2165	0.06668	1004.8	2604.3	1008.4	1795.7	2804.2	2.6457	6.1869	30.0
35.0	242.6	1.2347	0.05707	1045.4	2603.7	1049.8	1753.7	2803.4	2.7253	6.1353	35.0
40.0	250.4	1.1522	0.04978	1081.3	2602.3	1087.3	1714.1	2801.4	2.7964	6.0701	40.0
45.0	257.5	1.1692	0.04406	1116.2	2600.1	1121.9	1676.4	2798.3	2.8610	6.0199	45.0
50.0	264.0	1.2859	0.03944	1147.8	2597.1	1154.2	1640.1	2794.3	2.9202	5.9734	50.0
60.0	215.6	1.3187	0.03244	1205.4	2589.7	1213.4	1871.0	2784.3	3.0267	5.8892	60.0
70.0	285.9	1.3513	0.02737	1257.6	2580.5	1267.0	1505.1	2772.1	3.1211	5.8133	70.0
80.0	295.1	1.3842	0.02352	1305.6	2569.8	1316.6	1411.3	2758.0	3.2068	5.7432	80.0
90.0	303.4	1.4178	0.02048	1350.5	2557.8	1363.3	1378.9	2742.1	3.2858	5.6772	90.0
100.0	311.1	1.4524	0.01803	1393.0	2544.4	1407.6	1317.1	2724.7	3.3596	5.6141	100.0
110.0	318.2	1.4886	0.01599	1433.7	2529.8						100.0

TABLE A-3 (Continued)

Temp. °C	Press bar	Specific volume m³/kg Sat. Liquid $v_f \times 10^3$	Sat. Vapor v_s	Internal Energy kJ/kg Sat. Liquid u_f	Sat. Vapor u_g	Enthalpy kJ/kg Sat. Liquid h_f	Evap. h_{fg}	Sat. Vapor h_g	Entropy kJ/kg K Sat. Liquid s_f	Sat. Vapor s_g	Press bar
120.0	324.8	1.5267	0.01426	1473.0	2513.7	1450.1	1255.5	2705.6	3.4295	5.5527	110.0
130.0	330.9	1.5671	0.01278	1511.1	2496.1	1491.3	1193.6	2684.9	3.4962	5.4924	120.0
140.0	336.8	1.6107	0.01149	1548.6	2476.8	1531.5	1130.7	2662.2	3.5606	5.4323	130.0
150.0	342.2	1.6581	0.01034	1585.6	2455.5	1571.1	1066.5	2637.6	3.6232	5.3717	140.0
160.0	347.4	1.7107	0.009306	1622.7	2431.7	1610.5	1000.0	2610.5	3.6848	5.3098	150.0
170.0	352.4	1.7702	0.008364	1660.2	2405.0	1650.1	930.6	2580.6	3.7161	5.2455	160.0
180.0	357.1	1.8397	0.007489	1698.9	2374.3	1690.3	856.9	2547.2	3.8079	5.1777	170.0
190.0	361.5	1.9243	0.006657	1739.9	2338.1	1732.0	777.1	2509.1	3.8715	5.1044	180.0
200.0	365.8	1.036	0.005834	1785.6	2293.0	1776.5	688.0	2464.5	3.9388	5.0228	190.0
220.9	374.1	3.155	0.003155	2029.6	2029.6	1826.3	583.4	2409.7	4.0139	4.9269	200.0
						2099.3	0	2099.3	4.4298	4.4298	220.9

TABLE A-4 Properties of Superheated Water Vapor

T °C	p = 0.06 bar = 0.006 MPa (T_sat = 72.69°C)				p = 0.70 bar = 0.07 MPa (T_sat = 895°C)			
	v m³/kg	u kJ/kg	h kJ/kg	s kJ/kg K	v m³/kg	u kJ/kg	h kJ/kg	s kJ/kg K
Sat.	23.739	2425.0	2567.4	8.3304	2.365	2494.5	2660.0	7.4797
80	27.132	2487.3	2650.1	8.5804	2.434	2509.7	2680.0	7.5341
120	30.219	2544.7	2726.0	8.7840	2.571	2539.7	2719.6	7.6375
160	33.302	2602.7	2802.5	8.9693	2.841	2599.4	2798.2	7.8279
200	36.383	2661.4	2879.7	9.1398	3.108	2659.1	2876.7	8.0012
240	39.462	2721.0	2957.8	9.2982	3.374	2719.3	2955.5	8.1611
280	42.540	2781.5	3036.8	9.4464	3.640	2780.2	3035.0	8.3162
320	45.618	2843.0	3116.7	9.5859	3.905	2842.0	3115.3	8.4504
360	48.696	2905.5	3197.7	9.7180	4.170	2904.6	3196.5	8.5828
400	51.774	2969.0	3279.6	9.8435	4.434	2968.2	3278.6	8.786
440	54.851	3033.5	3362.6	9.9633	4.698	3031.9	3361.8	8.8286
500	59.467	3132.3	3489.1	10.1336	5.095	3131.8	3488.5	8.9991

T °C	p = 0.35 bar = 0.035 MPa (T_sat = 36.16°C)				p = 1.0 bar = 0.10 MPa (T_sat = 99.63°C)			
	v m³/kg	u kJ/kg	h kJ/kg	s kJ/kg K	v m³/kg	u kJ/kg	h kJ/kg	s kJ/kg K
Sat.	4.526	2473.0	2631.4	7.7158	1.694	2506.1	2675.5	7.3594
100	4.625	483.7	2645.6	7.7564	1.696	2506.7	2676.2	7.3614
120	5.163	2542.4	2723.1	7.9644	1.793	2537.3	2716.6	7.4668
160	5.696	2601.2	2800.6	8.1519	1.984	2597.8	2796.2	7.6597
200	6.228	2660.4	2878.4	8.3237	2.172	2658.1	2875.3	7.8343
240	6.758	2720.3	2956.8	8.4828	2.359	2718.5	2954.5	7.9949
280	7.287	2780.9	3036.0	8.6314	2.546	2779.6	3034.2	8.1445
320	7.815	2842.5	3116.1	8.7712	1.732	28415	3114.6	8.2849
360	8.344	2905.1	3197.1	8.9034	2.017	2904.2	3105.0	8.4175
400	8.872	2968.6	3279.2	9.0291	3.103	2967.9	3278.2	8.5435
440	9.400	3033.2	3362.2	9.1490	3.298	3032.6	3361.4	8.6636
500	10.192	3131.1	3488.8	9.3194	3.565	3131.6	3488.1	8.8342

TABLE A-4 (Continued)

p = 1.5 bar = 0.15 MPa (Tsat = 111.37°C)

T °C	v m³/kg	u kJ/kg	h kJ/kg	s kJ/kg K
Sat.	1.159	2519.7	2693.6	7.2183
120	1.188	2533.3	2711.4	7.2693
160	1.317	2595.2	2792.8	7.4665
200	1.444	2656.2	2872.9	7.6433
240	1.570	2717.2	2952.7	7.8052
280	1.695	2778.6	3032.8	7.9555
320	1.819	2840.6	3113.5	8.0964
360	1.943	2903.5	3195.0	8.2293
400	2.067	2967.3	3277.4	8.3555
440	2.191	3032.1	3360.7	8.4757
500	2.376	3131.2	3487.6	8.6466
600	2.685	3301.7	3704.3	8.9101

p = 0.06 bar = 0.006 MPa (Tsat = 72.69°C)

T °C	v m³/kg	u kJ/kg	h kJ/kg	s kJ/kg K
Sat.	0.3749	2361.2	2748.7	6.8213
180	0.4043	2609.7	2812.0	6.9636
200	0.4249	2642.9	2833.4	7.0392
240	0.4646	2707.6	2939.9	7.2307
280	0.5034	2771.2	3022.9	7.3865
320	0.3416	2834.7	3103.6	7.5308
360	0.3796	2898.7	3188.4	7.6660
400	0.173	2963.2	3279.9	7.7938
440	0.548	3028.6	3356.0	7.9132
500	0.7109	3138.4	3483.9	8.0873
600	0.3041	3299.6	3701.7	8.3322
700	0.8969	3477.5	3925.9	8.5952

p = 3.0 bar = 0.30 MPa (Tsat = 133.55°C)

T °C	v m³/kg	u kJ/kg	h kJ/kg	s kJ/kg K
Sat.	0.606	2543.6	2725.3	6.9919
160	0.651	2587.1	2782.3	7.1276
200	0.716	2650.7	2865.5	7.3115
240	0.781	2713.1	2947.3	7.4774
280	0.844	2778.4	3028.6	7.6299
320	0.907	2838.1	3110.1	7.7722
360	0.969	2901.4	3192.2	7.9061
400	1.032	2965.6	3275.0	8.0330
440	1.094	3030.6	3358.7	8.1538
500	1.187	3130.0	3486.0	8.3151
600	1.341	3300.8	3703.2	8.5892

p = 0.35 bar = 0.035 MPa (Tsat = 36.16°C)

T °C	v m³/kg	u kJ/kg	h kJ/kg	s kJ/kg K
Sat.	0.2729	2371.3	2763.5	6.7080
180	0.2847	2399.8	2799.1	6.7880
200	0.2999	2634.8	2844.8	6.8863
240	0.3292	2701.8	2932.2	7.0641
280	0.3574	2766.9	3017.1	7.2233
320	0.3832	2831.3	3100.9	7.3697
360	0.4126	2895.8	3184.7	7.3063
400	0.4397	2960.9	3268.7	7.6350
440	0.4667	3026.6	3333.3	7.7571
500	0.5070	3126.8	1481.7	7.9299
600	0.5738	3298.5	3700.2	8.1956
700	0.6403	3476.6	3924.8	8.4391

TABLE A-4 (Continued)

T °C	v m³/kg	u kJ/kg	h kJ/kg	s kJ/kg K
	p = 10.0 bar = 1.0 MPa (T_sat = 179.91°C)			
Sat.	0.1944	2383.6	2778.1	6.3865
200	0.2060	2621.9	2827.9	6.6940
240	0.2275	2692.9	2920.4	6.8817
280	0.2480	2760.2	3008.2	7.0465
320	0.2678	2826.1	3093.9	7.1962
360	0.2875	2891.6	3178.9	7.3349
400	0.3066	2937.3	3263.9	7.4631
440	0.3257	3023.6	3349.3	7.5883
500	0.3541	3124.4	3478.5	7.7622
540	0.3729	3192.6	3565.6	7.8720
600	0.4011	3296.8	3697.9	8.0290
640	0.4198	3367.4	3787.2	8.1290

T °C	v m³/kg	u kJ/kg	h kJ/kg	s kJ/kg K
	p = 20.0 bar = 2.0 MPa (T_sat = 212.42°C)			
Sat.	0.0996	2600.3	2799.5	6.3409
240	0.1085	2639.6	2876.5	6.4952
280	0.1200	2736.4	2976.4	6.6828
320	0.1308	2807.9	3069.5	6.8452
360	0.1411	2877.0	3159.3	6.9917
400	0.1512	2945.2	3247.6	7.1271
440	0.1611	3013.4	3335.5	7.2340
500	0.1757	3116.2	3467.8	7.4317
540	0.1853	3185.6	3556.1	7.5434
600	0.1996	3290.9	3690.1	7.7024
640	0.2091	3362.2	3780.4	7.8035
700	0.2232	3470.9	3917.4	7.9487

T °C	v m³/kg	u kJ/kg	h kJ/kg	s kJ/kg K
	p = 15.0 bar = 1.5 MPa (T_sat = 198.32°C)			
Sat.	0.1318	2394.5	2792.2	6.4448
200	0.1325	2598.1	2796.8	6.4546
240	0.1483	2676.9	2899.3	6.6628
280	0.1627	2748.6	2992.7	6.8381
320	0.1765	2817.1	3081.9	6.9938
360	0.1899	2884.4	3169.2	7.1363
400	0.2030	2951.3	3255.8	7.2690
440	0.2160	3018.3	3342.5	7.3940
500	0.2352	3120.3	3473.1	7.5698
540	0.2478	3189.1	3560.9	7.6803
600	0.2668	3393.9	3694.0	7.8385
640	0.2793	3364.8	3783.8	7.9391

T °C	v m³/kg	u kJ/kg	h kJ/kg	s kJ/kg K
	p = 30.0 bar = 3.0 MPa (T_sat = 233.90°C)			
Sat.	0.0667	2604.1	2804.2	6.1869
240	0.0682	2619.7	2824.3	6.2265
280	0.0771	2709.9	2941.3	6.4462
320	0.0850	2788.4	3043.4	6.6245
360	0.0923	2861.7	3138.7	6.7801
400	0.0994	2932.8	3230.9	6.9212
440	0.1062	3002.9	3321.5	7.0520
500	0.1162	3108.0	3436.5	7.2338
540	0.1227	3178.4	3546.6	7.3474
600	0.1324	3285.0	3682.3	7.5085
640	0.1388	3357.0	3773.5	7.6106
700	0.1484	3466.3	3911.7	7.7571

TABLE A-4 (Continued)

T °C	v m³/kg	u kJ/kg	h kJ/kg	s kJ/kg K	v m³/kg	u kJ/kg	h kJ/kg	s kJ/kg K
	$p = 40$ bar $= 4.0$ MPa ($T_{sat} = 250.04°C$)				$p = 60$ bar $= 6.0$ MPa ($T_{sat} = 275.64°C$)			
Sat.	0.04978	2602.3	2801.4	6.0701	0.03244	2589.7	2784.3	5.8892
280	0.05546	2680.0	2901.8	6.2568	0.03317	2605.2	2804.2	5.9252
320	0.06199	2767.4	3015.4	6.4553	0.03876	2720.0	2952.6	6.1846
360	0.06788	2845.7	3117.2	6.6215	0.04331	2811.5	3071.1	6.3782
400	0.07341	2919.9	3213.6	6.7690	0.04739	2892.9	3177.2	6.5408
440	0.07872	2992.2	3307.1	6.9041	0.05122	2970.0	3277.3	6.6853
500	0.08643	3099.5	3445.3	7.0901	0.05665	3082.2	3422.2	6.8803
540	0.09145	3171.1	3536.9	7.2056	0.06015	3156.1	3517.0	6.9999
600	0.09885	3279.1	3674.4	7.3688	0.06525	3266.9	3658.4	7.1677
640	0.1037	3351.8	3766.6	7.4720	0.06859	3341.0	3752.6	7.2731
700	0.1110	3461.1	3905.9	7.6198	0.07352	3453.1	3894.1	7.4234
740	0.1157	3536.6	3999.6	7.7141	0.07677	3523.3	3989.2	7.5190
	$p = 80$ bar $= 8.0$ MPa ($T_{sat} = 295.06°C$)				$p = 100$ bar $= 10.0$ MPa ($T_{sat} = 311.06°C$)			
Sat.	0.02332	2569.8	2758.0	5.7432	0.01803	2544.4	2724.7	5.6141
320	0.02682	2662.7	2877.2	5.9489	0.01925	2588.8	2781.3	5.7103
360	0.03089	2772.7	3019.8	6.1819	0.02331	2729.1	2962.1	6.0060
400	0.03432	2863.8	3138.3	6.3634	0.02641	2832.4	3096.5	6.2120
440	0.03742	2946.7	3246.1	6.5190	0.02911	2922.1	3213.2	6.3805
480	0.04034	3025.7	3348.4	6.6586	0.03160	3005.4	3321.4	6.5282
520	0.04313	3102.7	3447.7	6.7871	0.03394	3085.6	3425.1	6.6622
560	0.04582	3178.7	3545.3	6.9072	0.03619	3164.1	3526.0	6.7864
600	0.04845	3254.4	3642.0	7.0206	0.03837	3241.7	3625.3	6.9029
640	0.05102	3330.1	3738.3	7.1283	0.04048	3313.9	3723.7	7.0131
700	0.05481	3443.9	3882.4	7.2812	0.04358	3434.7	3870.5	7.1687
740	0.05729	3520.4	3978.7	7.3732	0.04560	3512.1	3968.1	7.2670

TABLE A-4 (Continued)

T °C	v m³/kg	u kJ/kg	h kJ/kg	s kJ/kg K	v m³/kg	u kJ/kg	h kJ/kg	s kJ/kg K
	p = 120 bar = 12.0 MPa (T_sat = 324.75°C)				p = 140 bar = 14.0 MPa (T_sat = 336.75°C)			
Sat.	0.01426	2513.7	2684.9	5.4924	0.01149	2476.8	2637.6	5.3717
360	0.01811	2678.4	2895.7	5.8361	0.01422	2617.4	2816.5	5.6602
400	0.02108	2798.3	3051.3	6.0747	0.01722	2760.9	3001.9	5.9448
440	0.02355	2896.1	3178.7	6.2586	0.01954	2868.6	3142.2	6.1474
480	0.02376	2954.4	3293.5	6.4154	0.02157	2962.5	3264.5	6.3143
520	0.02781	3068.0	3401.5	6.5555	0.02343	3049.8	3377.8	6.4610
560	0.02977	3149.0	506.2	6.6840	0.02517	3133.6	3486.0	6.5941
600	0.03164	3228.7	3608.3	6.8037	0.02683	3215.4	3591.1	6.7172
640	0.03345	3307.5	3709.0	6.9164	0.02843	3296.0	3694.1	6.8326
700	0.03610	3425.2	3858.4	7.0749	0.03075	3415.7	3846.2	6.9939
740	0.03781	3503.7	3957.4	7.1746	0.03225	3495.2	3946.7	7.0952

T °C	v m³/kg	u kJ/kg	h kJ/kg	s kJ/kg K	v m³/kg	u kJ/kg	h kJ/kg	s kJ/kg K
	p = 160 bar = 16.0 MPa (T_sat = 347.44°C)				p = 180 bar = 18.0 MPa (T_sat = 357.06°C)			
Sat.	0.00931	2431.7	2580.6	5.2455	0.00749	2374.3	2509.1	5.1044
360	0.01105	2539.0	2715.8	5.4614	0.00809	2418.9	2564.5	5.1922
400	0.01426	2719.4	2947.6	5.8175	0.01190	2672.8	2887.0	5.6887
440	0.01652	2839.4	3103.7	6.0420	0.01414	2808.2	3062.8	5.4428
480	0.01842	2939.7	3234.4	6.2215	0.01596	2915.9	3203.2	6.1345
520	0.02013	3031.1	3353.3	6.3752	0.01757	3011.8	3378.0	6.2960
560	0.02172	3117.8	3463.4	6.5132	0.01904	3101.7	3444.4	6.4392
600	0.012323	3201.8	3573.5	6.6399	0.02042	3188.0	3335.6	6.5696
640	0.00467	3284.2	3678.9	6.7580	0.02174	3272.3	3663.6	6.6905
700	0.02674	3406.0	3833.9	6.9224	0.02362	3396.3	3821.5	6.8580
740	0.02808	3486.7	3935.9	7.0251	0.02483	3478.0	3925.0	6.9623

TABLE A-4 (Continued)

T °C	p = 200 bar = 20.0 MPa (T_sat = 365.81°C)				p = 240 bar = 24.0 MPa			
	v m³/kg	u kJ/kg	h kJ/kg	s kJ/kg K	v m³/kg	u kJ/kg	h kJ/kg	s kJ/kg K
Sat.	0.00583	2293.0	2409.7	4.9269				
400	0.00994	2619.3	2818.1	5.5540	0.00673	2477.8	2639.4	5.2393
440	0.011222	2774.9	3019.4	5.8450	0.00929	2700.6	2923.4	5.6506
480	0.01399	2891.2	3170.8	6.0518	0.01100	2838.3	3102.3	5.8950
520	0.01551	2992.0	3302.2	6.2218	0.01241	2950.5	3248.5	6.0842
560	0.01689	3085.2	3423.0	6.3705	0.01366	3051.1	3379.0	6.2448
600	0.01818	3174.0	3537.6	6.5048	0.01481	3145.2	3500.7	6.3875
640	0.01940	3260.2	3648.1	6.6286	0.01588	3235.3	3616.7	6.3174
700	0.02113	3386.4	3809.0	6.7993	0.01739	3366.4	3783.8	6.6947
740	0.02224	3469.3	3914.1	6.9052	0.01835	3451.7	3892.1	6.8038
800	0.02385	3592.7	4069.7	7.0544	0.01974	3578.0	4051.6	6.9567

T °C	p = 280 bar = 28.0 MPa				p = 320 bar = 32A MPa			
	v m³/kg	u kJ/kg	h kJ/kg	s kJ/kg K	v m³/kg	u kJ/kg	h kJ/kg	s kJ/kg K
400	0.00383	2223.5	2330.7	4.7494	0.00236	1980.4	2055.9	4.3239
440	0.00712	2611.2	2812.6	5.4494	0.00544	2509.0	2683.0	5.2327
480	0.00885	2780.8	3028.5	5.7446	0.00722	2718.1	2949.2	5.5968
520	0.01020	2906.8	3192.3	5.9566	0.00833	2860.7	3133.7	5.8357
560	0.01136	3015.7	3333.7	6.1307	0.00963	2979.0	3257.2	6.0246
600	0.01241	3115.6	3463.0	6.2823	0.01061	3085.3	3424.6	6.1858
640	0.01338	3210.3	3584.8	6.4187	0.01150	3184.5	3582.5	6.3290
700	0.01473	3346.1	3758.4	6.6029	0.01271	3325.4	3732.8	6.5201
740	0.01558	3433.9	3871.0	6.7153	0.01350	3415.9	3847.8	6.6361
800	0.01680	3563.1	4033.4	6.8720	0.01460	3548.0	4015.1	6.7966
900	0.01873	3774.3	4298.8	7.1084	0.01633	3762.7	4285.1	7.0372

TABLE A-5 Properties of Compressed Liquid Water

T °C	v × 10³ m³/kg	u kJ/kg	h kJ/kg	s kJ/kg K	v × 10³ m³/kg	u kJ/kg	h kJ/kg	s kJ/kg K
	p = 25 bar = 2.5 MPa (T_{sat} = 223.90°C)				p = 50 bar = 5.0 MPa (T_{sat} = 263.99°C)			
20	1.0006	83.80	86.30	0.2961	0.9995	83.45	88.165	0.2956
40	1.0067	167.25	169.77	0.5713	1.0056	166.95	171.97	0.5705
80	1.0280	334.29	336.56	1.0737	1.0268	333.72	338.85	1.0720
100	1.0423	418.24	420.85	1.3050	1.0410	417.52	422.72	1.3030
140	1.0784	587.82	590.51	1.7369	1.0768	586.76	592.15	1.7343
180	1.1261	761.16	763.97	2.1375	1.1140	759.63	765.25	2.1341
200	1.1555	849.9	852.8	2.3294	1.1530	848.1	853.9	2.3255
220	1.1898	940.7	943.7	2.5174	1.1866	938.4	944.4	2.5128
Sat.	1.1973	959.1	962.1	2.5546	1.2859	1147.8	1154.2	2.9202

T °C	v × 10³ m³/kg	u kJ/kg	h kJ/kg	s kJ/kg K	v × 10³ m³/kg	u kJ/kg	h kJ/kg	s kJ/kg K
	p = 75 bar = 7.5 MPa (T_{sat} = 290.59°C)				p = 100 bar = 10.0 MPa (T_{sat} = 311.06°C)			
20	0.9984	83.50	90.99	0.2950	0.9972	83.36	93.33	0.2945
40	1.0045	166.64	174.18	0.5696	1.0034	166.35	176.38	0.5686
80	1.0256	333.15	340.84	1.0704	1.0245	332.59	342.83	1.0688
100	1.0397	416.81	424.62	1.3011	1.0385	416.12	426.50	1.2992
140	1.0752	585.72	593.78	1.7317	1.0737	584.68	595.42	1.7292
180	1.1219	758.13	766.55	2.1308	1.1199	756.65	767.84	2.1275
220	1.1835	936.2	945.1	2.5083	1.1805	934.1	945.9	2.5039
260	1.2696	1124.4	1134.0	2.8763	1.2645	1121.1	1133.7	2.8699
Sat.	1.3677	1382.0	1292.2	3.1649	1.4524	1393.0	1407.6	3.3596

TABLE A-5 (Continued)

T °C	v × 10³ m³/kg	u kJ/kg	h kJ/kg	s kJ/kg K	v × 10³ m³/kg	u kJ/kg	h kJ/kg	s kJ/kg K
	p = 150 bar = 15.0 MPa (T_sat = 342.24°C)				p = 200 bar = 20.0 MPa (T_sat = 365.8°C)			
20	.9950	83.06	97.99	0.2934	0.9928	82.77	102.62	0.2923
40	1.0013	165.76	180.78	0.5666	0.9992	165.17	185.16	0.5646
80	1.0222	331.48	346.81	1.0656	1.0199	330.40	350.80	1.0624
100	1.0361	414.74	430.28	1.2955	1.0337	413.39	434.06	1.2917
140	1.0707	582.66	598.72	1.7242	1.0675	580.69	602.04	1.7193
180	1.1159	753.76	770.50	2.1210	1.1120	750.95	773.20	2.1147
220	1.1748	929.9	947.5	2.4953	1.1693	925.9	949.3	2.4870
260	1.2550	1114.6	1133.4	2.8576	1.2462	1108.6	1133.5	2.8459
300	1.3770	1316.6	1337.3	3.2260	1.3596	1306.1	1333.3	3.2071
Sat.	1.6581	1585.6	1610.5	3.6848	2.036	1785.6	1626.3	4.0139
	p = 250 bar = 25 MPa				p = 300 bar = 30.0 MPa			
20	0.9907	82.47	107.24	0.2911	0.9886	82.17	111.84	0.2899
40	0.9971	164.60	189.52	0.5626	0.9951	164.04	193.89	0.5607
100	1.0313	412.08	437.85	1.2881	1.0290	410.78	441.66	1.2844
200	1.1344	834.5	862.8	2.2961	1.1302	831.4	865.3	2.2893
300	1.3442	1296.6	1330.2	3.1900	1.3304	1287.9	1327.8	3.1741

TABLE A-6 Properties of Saturated Water (Solid-Vapor): Temperature Table

Temp. °C	Pressure kPa	Specific volume m³/kg		Internal Energy kJ/kg			Enthalpy kJ/kg			Entropy kJ/kg K		
		Sat. Solid $v_1 \times 10^3$	Sat. Vapor v_g	Sat. Solid u_i	Subl. u_{ig}	Sat. Vapor u_g	Sat. Solid h_i	Subl. h_{ig}	Sat. Vapor h_g	Sat. Solid s_i	Subl. s_{ig}	Sat. Vapor s_g
0.01	0.6113	1.0908	206.1	−333.40	2708.7	2375.3	−33.40	2834.8	2501.4	−1.221	10.378	9.156
0	0.6108	1.0908	206.3	−333.43	2708.8	2375.3	−333.43	2834.8	1501.3	−1.221	10.378	9.157
−2	0.5176	1.0904	241.7	−337.62	2710.2	2372.6	−337.62	2835.3	2497.7	−1.237	10.456	9.219
−4	0.4573	1.0901	283.8	−341.78	2711.6	2369.8	−341.78	2835.7	2494.0	−1.253	10.536	9.283
−6	0.3689	1.0898	334.2	−345.91	2712.9	2367.0	−345.91	2836.2	2490.3	−1.268	10.616	9.348
−8	0.3102	1.0894	394.4	−350.02	2714.2	2364.2	−350.02	2836.6	2486.6	−1.284	10.698	9.414
−10	0.2602	1.0891	466.7	−354.09	2715.5	2361.4	−354.09	2837.0	2482.9	−1.299	10.781	9.481
−12	0.2176	1.0888	553.7	−358.14	2716.8	2358.7	−358.14	2837.3	2479.2	−1.313	10.865	9.550
−14	0.1813	1.0884	658.8	−362.15	2718.0	2355.9	−362.15	2837.6	2475.5	−1.331	10.950	9.619
−16	0.1510	1.0881	756.0	−366.14	2719.2	2353.1	−366.14	2837.9	2471.8	−1.346	11.036	9.690
−18	0.1252	1.0878	940.5	−370.10	2720.4	2350.3	−370.10	2838.2	2468.1	−1.362	11.123	9.762
−20	0.1035	1.0874	1128.6	−374.03	2721.6	2347.5	−374.03	2838.4	2464.3	−1.377	11.212	9.835
−22	0.0853	1.0871	1358.4	−377.93	2722.7	2344.7	−377.93	2838.6	2460.6	−1.393	11.302	9.909
−24	0.0701	1.0868	1640.1	−381.80	2723.7	2342.0	−381.80	2838.7	2456.9	−1.408	11.394	9.985
−26	0.0574	1.0864	1986.4	−385.64	2724.8	2339.2	−385.64	2838.9	2453.2	−1.424	11.486	10.062
−28	0.0469	1.0861	2413.7	−389.45	2725.8	2336.4	−389.45	2839.0	2449.5	−1.439	11.580	10.141
−30	0.0381	1.0858	2943	−393.23	2726.8	2333.6	−393.23	2839.0	2445.8	−1.455	11.676	10.221
−32	0.0309	1.0854	3600	−396.98	2727.8	2330.8	−396.98	2839.1	2442.1	−1.471	11.773	10.303
−34	0.0250	1.0851	4419	−400.71	2728.7	2328.0	−400.71	2839.1	2438.4	−1.486	11.872	10.386
−36	0.0201	1.0848	5444	−404.40	2729.6	2325.2	−404.40	2839.1	2434.7	−1.501	11.972	10.470
−38	0.0161	1.0844	6731	−408.06	2730.5	2322.4	−408.06	2839.0	2430.9	−1.517	12.073	10.556
−40	0.0129	1.0841	8354	−411.70	2731.3	2319.6	−411.70	2838.9	2427.2	−1.532	12.176	10.644

TABLE A-7 Properties of Saturated Refrigerant 22 (Liquid-Vapor): Temperature Table

Temp. °C	Press bar	Specific volume m³/kg		Internal Energy kJ/kg		Enthalpy kJ/kg			Entropy kJ/kg K		Temp °C
		Sat. Liquid $v_f \times 10^3$	Sat. Vapor v_g	Sat. Liquid u_f	Sat. Vapor u_g	Sat. Liquid h_f	Evap. h_{fg}	Sat. Vapor h_g	Sat. Liquid s_f	Sat. Vapor s_g	
−60	0.3749	0.6833	0.5370	−21.57	203.67	−21.55	245.35	223.81	−0.0964	1.0547	−60
−50	0.6451	0.6966	0.3239	−10.89	207.70	−10.85	239.44	228.60	−0.0474	1.0256	−50
−45	0.8290	0.7037	0.2564	−5.50	209.70	−5.44	236.39	230.95	−0.0235	1.0126	−45
−40	1.0522	0.7109	0.2052	−0.07	211.68	0.00	233.27	233.27	0.0000	1.0005	−40
−36	1.2627	0.7169	0.1730	4.29	213.25	4.38	230.71	235.09	0.0186	0.9914	−36
−32	1.5049	0.7231	0.1468	8.68	214.80	8.79	228.10	236.89	0.0369	0.9828	−32
−30	1.6389	0.7262	0.1355	10.88	215.58	11.00	226.77	237.78	0.0460	0.9787	−30
−28	1.7819	0.7294	0.1252	13.09	216.34	13.22	225.43	238.66	0.0551	0.9746	−28
−26	1.9345	0.7327	0.1159	15.31	217.11	15.45	224.08	239.53	0.0641	0.9707	−26
−22	2.2695	0.7393	0.0997	19.76	218.62	19.92	221.32	241.24	0.0819	0.9631	−22
−20	2.4534	0.7427	0.0926	21.99	219.37	22.17	219.91	242.09	0.0905	0.9595	−20
−18	2.6482	0.7462	0.0861	24.23	220.11	24.43	218.49	242.92	0.0996	0.9559	−18
−16	2.5547	0.7497	0.0802	26.48	220.85	26.69	217.05	243.74	0.1084	0.9525	−16
−14	3.0733	0.7533	0.0748	28.73	221.58	28.97	215.59	244.56	0.1171	0.9490	−14
−12	3.3044	0.7569	0.0698	31.00	222.30	31.25	214.11	245.36	0.1255	0.9457	−12
−10	3.5485	0.7606	0.0652	33.27	223.02	33.54	212.62	246.15	0.1345	0.9424	−10
−8	3.8062	0.7644	0.0610	35.54	223.73	35.83	211.10	246.93	0.1431	0.9392	−8
−6	4.0777	0.7683	0.0571	37.83	224.43	38.14	209.56	247.70	0.1517	0.9361	−6
−4	4.3638	0.7722	0.0535	40.12	225.13	40.46	208.00	248.45	0.1602	0.9330	−4
−2	4.6647	0.7762	0.0501	42.42	225.82	42.78	206.41	249.20	0.1688	0.9300	−2
0	4.9811	0.7803	0.0470	44.73	226.50	45.12	204.81	249.92	0.1773	0.9271	0
2	5.3133	0.7844	0.0442	47.04	227.17	47.46	203.18	250.64	0.1857	0.9241	2
4	5.6619	0.7887	0.0415	49.37	227.83	49.82	201.52	251.34	0.1941	0.9213	4

TABLE A-7 (Continued)

Temp. °C	Press bar	Specific volume m³/kg		Internal Energy kJ/kg		Enthalpy kJ/kg			Entropy kJ/kg K		Temp °C
		Sat. Liquid $v_f \times 10^3$	Sat. Vapor v_g	Sat. Liquid u_f	Sat. Vapor u_g	Sat. Liquid h_f	Evap. h_g	Sat. Vapor h_g	Sat. Liquid s_f	Sat. Vapor s_g	
6	6.0275	0.7930	0.0391	51.71	228.48	52.18	199.84	252.03	0.2025	0.9184	6
8	6.4105	0.7974	0.0368	54.05	229.13	54.56	198.14	252.70	0.2109	0.9157	8
10	6.8113	0.8020	0.0346	56.40	229.76	56.95	196.40	253.35	0.2193	0.9129	10
12	7.2307	0.8066	0.0326	58.77	230.38	59.35	194.64	253.99	0.2276	0.9102	12
16	8.1265	0.8162	0.0291	63.53	231.59	64.19	191.02	255.21	0.2442	0.9048	16
20	9.1030	0.8263	0.0259	68.33	232.76	69.09	187.28	256.37	0.2607	0.5996	20
24	10.164	0.8369	0.0232	73.19	233.87	74.04	183.40	257.44	0.2772	0.8944	24
28	11.313	0.8480	0.0208	78.09	234.92	79.05	179.37	258.43	0.2936	0.8893	28
32	12.556	0.8599	0.0186	83.06	235.91	84.14	175.18	259.32	0.3101	0.8842	32
36	13.897	0.8724	0.0168	88.08	236.83	89.29	170.82	260.11	0.3265	0.8790	36
40	15.341	0.8858	0.0151	93.18	237.66	94.53	166.25	260.79	0.3429	0.8738	40
45	17.298	0.9039	0.0132	99.65	238.59	101.21	160.24	261.46	0.3635	0.8672	45
50	19.433	0.9238	0.0116	106.26	239.34	108.06	153.84	261.90	0.3842	0.8603	50
60	24.281	0.9705	0.0089	120.00	240.24	122.35	139.61	261.96	0.4264	0.8455	60

TABLE A-8 Properties of Saturated Refrigerant 22 (Liquid-Vapor): Pressure Table

Temp. °C	Press bar	Specific volume m³/kg		Internal Energy kJ/kg		Enthalpy kJ/kg			Entropy kJ/kg K		Temp °C
		Sat. Liquid $v_f \times 10^3$	Sat. Vapor v_g	Sat. Liquid u_f	Sat. Vapor u_g	Sat. Liquid h_f	Evap. h_g	Sat. Vapor h_g	Sat. Liquid s_f	Sat. Vapor s_g	
0.40	-58.86	0.6847	0.5056	-20.36	204.13	-20.34	244.69	224.36	-0.0907	1.0512	0.40
0.50	-54.83	0.6901	0.4107	-16.07	205.76	-16.03	242.33	226.30	-0.0709	1.0391	0.50
0.60	-51.40	0.6947	0.3466	-12.39	207.14	-12.35	240.28	227.93	-0.0542	1.0294	0.60
0.70	-48.40	0.6989	0.3002	-9.17	208.34	-9.12	238.47	229.35	-0.0397	1.0213	0.70
0.80	-45.73	0.7026	0.2650	-6.28	209.41	-6.23	236.84	230.61	-0.0270	1.0144	0.80
0.90	-43.30	0.7061	0.2374	-3.66	210.37	-3.60	235.34	231.74	-0.0155	1.0084	0.90
1.00	-41.09	0.7093	0.2152	-1.26	211.25	-1.19	233.95	232.77	-0.0051	1.0031	1.00
1.25	-36.23	0.7166	0.1746	4.04	213.16	4.13	230.86	234.99	0.0175	0.9919	1.25
1.50	-32.08	0.7230	0.1472	8.60	214.77	8.70	228.15	236.86	0.0366	0.9830	1.50
1.75	-28.44	0.7287	0.1274	12.61	216.18	12.74	225.73	238.47	0.0531	0.9755	1.75
2.00	-25.18	0.7340	0.1123	16.22	217.42	16.37	223.52	239.88	0.0678	0.9691	2.00
2.25	-22.22	0.7389	0.1005	19.51	218.53	19.67	221.47	241.15	0.0809	0.9636	2.25
2.50	-19.51	0.7436	0.0910	22.54	219.55	22.72	219.57	242.29	0.0930	0.9586	2.50
2.75	-17.00	0.7479	0.0831	25.36	220.48	25.56	217.77	243.33	0.1040	0.9542	2.75
3.00	-14.66	0.7521	0.0765	27.99	221.34	28.22	216.07	244.29	0.1143	0.9502	3.00
3.25	-12.46	0.7561	0.0709	30.47	222.13	30.72	214.46	245.18	0.1238	0.9465	3.25
3.50	-10.39	0.7599	0.0661	32.82	222.88	33.09	212.91	246.00	0.1328	0.9431	3.50
3.75	-8.43	0.7636	0.0618	35.06	223.58	35.34	211.42	246.77	0.1413	0.9399	3.75
4.00	-6.56	0.7672	0.0581	37.18	224.24	37.49	209.99	247.48	0.1493	0.9370	4.00
4.25	-4.78	0.7706	0.0548	39.22	224.86	39.55	208.61	248.16	0.1569	0.9342	4.25
4.50	-3.08	0.7740	0.0519	41.17	225.45	41.52	207.27	248.80	0.1642	0.9316	4.50
4.75	-1.45	0.7773	0.0492	43.05	226.00	43.42	205.98	249.40	0.1711	0.9292	4.75
5.00	0.12	0.7805	0.0469	44.86	226.54	45.25	204.71	249.97	0.1777	0.9269	5.00
5.25	1.63	0.7836	0.0447	46.61	227.04	47.02	203.48	250.51	0.1841	0.9247	5.25

TABLE A-8 (Continued)

Temp. °C	Press bar	Specific volume m³/kg		Internal Energy kJ/kg		Enthalpy kJ/kg			Entropy kJ/kg K		Temp °C
		Sat. Liquid $v_f \times 10^3$	Sat. Vapor v_g	Sat. Liquid u_f	Sat. Vapor u_g	Sat. Liquid h_f	Evap. h_g	Sat. Vapor h_g	Sat. Liquid s_f	Sat. Vapor s_g	
5.50	3.08	0.7867	0.0427	48.30	227.53	48.74	202.28	251.02	0.1903	0.9226	5.50
5.75	4.49	0.7897	0.0409	49.94	227.99	50.40	201.11	251.51	0.1962	0.9206	5.75
6.00	5.85	0.7927	0.0392	51.53	228.44	52.01	199.97	251.98	0.2019	0.9186	6.00
7.00	10.91	0.8041	0.0337	57.48	230.04	58.04	195.60	253.64	0.2231	0.9117	7.00
8.00	15.45	0.8149	0.0295	62.88	231.43	63.53	191.52	255.05	0.2419	0.9056	8.00
9.00	19.59	0.8252	0.0262	67.84	232.64	68.59	187.67	256.25	0.2591	0.9001	9.00
10.00	23.40	0.8352	0.0236	72.46	233.71	73.30	183.99	257.28	0.2748	0.8952	10.00
12.00	30.25	0.8546	0.0195	80.87	235.48	81.90	177.04	258.94	0.3029	0.8864	12.00
14.00	36.29	0.8734	0.0166	88.45	236.89	89.68	170.49	260.16	0.3277	0.8786	14.00
16.00	41.73	0.8919	0.0144	95.41	238.00	96.83	164.21	261.04	0.3500	0.8715	16.00
18.00	46.69	0.9104	0.0127	101.87	238.86	103.51	158.13	261.64	0.3705	0.8649	18.00
20.00	51.26	0.9291	0.0112	107.95	239.51	109.81	152.17	261.98	0.3895	0.8586	20.00
24.00	59.46	0.9677	0.0091	119.24	240.22	121.56	140.43	261.99	0.4241	0.8463	24.00

TABLE A-9 Properties of Superheated Refrigerant 22 Vapor

$p = 0.4$ bar $= 0.04$ MPa ($T_{sat} = -58.86°C$)

T °C	$v \times 10^3$ m³/kg	u kJ/kg	h kJ/kg	s kJ/kg·K
Sat.	0.50559	204.13	224.36	1.0512
−55	0.51532	205.92	226.53	1.0612
−50	0.52787	208.26	229.38	1.0741
−45	0.54037	210.63	232.24	1.0868
−40	0.55234	213.02	235.13	1.0993
−35	0.56526	215.43	238.05	1.1117
−30	0.57766	217.88	240.99	1.1239
−25	0.59002	220.35	243.95	1.1360
−20	0.60236	222.85	246.95	1.1479
−15	0.61468	225.38	249.97	1.1597
−10	0.62697	227.93	253.01	1.1714
−5	0.63925	230.52	256.09	1.1830
0	0.65151	233.13	259.19	1.1944

$p = 0.6$ bar $= 0.06$ MPa ($T_{sat} = -51.40°C$)

T °C	$v \times 10^3$ m³/kg	u kJ/kg	h kJ/kg	s kJ/kg·K
Sat.	0.34656	207.14	227.93	1.0294
−50	0.34595	207.80	228.74	1.0330
−45	0.35747	210.20	231.65	1.0459
−40	0.36594	212.62	234.58	1.0586
−35	0.37437	215.06	237.52	1.0711
−30	0.38277	217.53	240.49	1.0835
−25	0.39114	220.02	243.49	1.0956
−20	0.39948	222.54	246.51	1.1077
−15	0.40779	225.08	249.55	1.1196
−10	0.41608	227.65	252.62	1.1314
−5	0.42436	230.25	255.71	1.1430
0	0.43261	232.88	258.83	1.1545

$p = 0.8$ bar $= 0.08$ MPa ($T_{sat} = -45.73°C$)

T °C	$v \times 10^3$ m³/kg	u kJ/kg	h kJ/kg	s kJ/kg·K
Sat.	0.26503	209.41	230.61	1.0144
−45	0.26597	209.76	231.04	1.0163
−40	0.27245	212.21	234.01	1.0292
−35	0.27590	214.68	236.99	1.0418
−30	0.28530	217.17	239.99	1.0543
−25	0.29167	219.68	243.02	1.0666
−20	0.29801	222.22	246.06	1.0788
−15	0.30433	224.78	249.13	1.0908
−10	0.31062	227.37	252.22	1.1026

$p = 1.0$ bar $= 0.10$ MPa ($T_{sat} = -41.09°C$)

T °C	$v \times 10^3$ m³/kg	u kJ/kg	h kJ/kg	s kJ/kg·K
Sat.	0.21518	211.25	232.77	1.0031
−45	0.21633	211.79	233.42	1.0059
−40	0.22158	214.29	236.44	1.0187
−35	0.22679	216.80	239.48	1.0313
−30	0.23197	219.34	242.54	1.0438
−25	0.23712	221.90	245.61	1.0560
−20	0.24224	224.48	248.70	1.0681
−15	0.24734	227.08	251.82	1.0801
−10	0.25241	229.71	254.95	1.0919

TABLE A-9 (Continued)

T °C	v × 10³ m³/kg	u kJ/kg	h kJ/kg	s kJ/kg K	v × 10³ m³/kg	u kJ/kg	h kJ/kg	s kJ/kg K
−5	0.31690	229.98	255.34	1.1143	0.25747	232.36	258.11	1.1035
0	0.32315	232.62	258.47	1.1259	0.26251	235.04	261.29	1.1151
5	0.32939	235.29	261.64	1.1374	0.26753	237.74	264.50	1.1265
10	0.33561	237.98	264.83	3.1488				
	p = 1.5 bar = 0.15 MPa (T_{sat} = −32.08°C)				p = 2.0 bar = 0.20 MPa (T_{sat} = −25.18°C)			
Sat.	0.14721	214.77	236.86	0.9830	0.11232	217.42	239.88	0.9691
−30	0.14872	215.85	238.16	0.9883				
−25	0.15232	218.45	241.30	1.0011	0.11242	217.51	240.00	0.9696
−20	0.15588	221.07	244.45	1.0137	0.11520	220.19	243.23	0.9825
−15	0.15941	223.70	247.61	1.0260	0.11795	222.88	246.47	0.9952
−10	0.16292	226.35	250.78	1.0382	0.12067	225.58	249.72	1.0076
−5	0.16640	229.02	253.98	1.0502	0.12336	228.30	252.97	1.0199
0	0.16957	231.70	257.18	1.0621	0.12603	231.03	256.23	1.0310
5	0.17331	234.42	260.41	1.0738	0.12868	233.78	259.51	1.0438
10	0.17674	237.15	263.66	1.0854	0.13132	236.54	262.81	1.0555
15	0.18015	239.91	266.93	1.0968	0.13393	239.33	266.12	1.0671
20	0.18355	242.69	270.22	1.1081	0.13653	242.14	269.44	1.0786
25	0.18693	245.49	273.53	1.1193	0.13912	244.97	272.79	1.0899
	p = 2.5 bar = 0.25 MPa (T_{sat} = −19.51°C)				p = 3.0 bar = 0.30 MPa (T_{sat} = −14.66°C)			
Sat.	0.09097	219.55	242.29	0.9586	0.07651	221.34	244.29	0.9502
−15	0.09303	222.03	245.29	0.9703	0.07833	223.96	247.46	0.9623
−10	0.09528	224.79	248.61	0.9831	0.08025	226.78	250.86	0.9751
−5	0.09751	227.55	251.93	0.9956	0.08214	229.61	254.25	0.9876
0	0.09971	230.33	255.26	1.0078	0.08400	232.44	257.64	0.9999

TABLE A-9 (Continued)

T °C	v×10³ m³/kg	u kJ/kg	h kJ/kg	s kJ/kg K	v×10³ m³/kg	u kJ/kg	h kJ/kg	s kJ/kg K
5	0.10189	233.12	258.59	1.0199				
10	0.10405	235.92	261.93	1.0318	0.08585	235.28	261.04	1.0120
15	0.10619	238.74	265.29	1.0436	0.08767	238.14	264.44	1.0239
20	0.10831	241.58	268.66	1.0552	0.08949	241.01	267.85	1.0357
25	0.11043	244.44	272.04	1.0666	0.09128	243.89	271.28	1.0472
30	0.11253	247.31	275.44	1.0779	0.09307	246.80	274.72	1.0537
35	0.11461	250.21	278.86	1.0891	0.09484	249.72	278.17	1.0700
40	0.11669	253.13	282.30	1.1002	0.09660	252.66	281.64	1.0811
	$p = 3.5$ bar $= 0.35$ MPa ($T_{sat} = -10.39°C$)				$p = 4.0$ bar $= 0.40$ MPa ($T_{sat} = -6.56°C$)			
Sat.	0.06605	222.88	246.00	0.9431	0.05812	224.24	247.48	0.9370
−10	0.06619	223.10	246.27	0.9441	0.05860	225.36	248.60	0.9411
−5	0.06789	225.99	249.75	0.9572	0.06011	228.09	252.14	0.9542
0	0.06956	228.86	253.21	0.9700	0.06160	231.02	225.66	0.9670
5	0.07121	231.74	256.67	0.9825	0.06306	233.95	259.18	0.9795
10	0.07284	234.63	260.12	0.9948	0.06450	236.89	262.69	0.9918
15	0.07444	237.52	263.57	1.0069	0.06592	239.83	266.19	1.0039
20	0.07603	240.42	267.03	1.0188	0.06733	242.77	269.71	1.0153
25	0.07760	243.34	270.50	1.0305	0.06872	245.73	273.22	1.0274
30	0.07916	246.27	273.97	1.0421	0.07010	248.71	276.75	1.0390
35	0.08070	249.22	227.46	1.0535	0.07146	251.70	280.28	1.0504
40	0.08224	252.18	280.97	1.0648	0.07282	254.70	283.83	1.0616
45	0.08376	255.17	284.48	1.0759				

TABLE A-9 (Continued)

T °C	$v \times 10^3$ m³/kg	u kJ/kg	h kJ/kg	s kJ/kg K	$v \times 10^3$ m³/kg	u kJ/kg	h kJ/kg	s kJ/kg K
	p = 4.5 bar = 0.45 MPa (T_{sat} = −3.08°C)				p = 5.0 bar = 0.50 MPa (T_{sat} = 0.12°C)			
Sat.	0.05189	225.45	248.80	0.9316	0.04686	226.54	249.97	0.9269
0	0.05275	227.29	251.03	0.9399				
5	0.05411	230.28	254.63	0.9529	0.04810	229.52	253.57	0.9399
10	0.05545	233.26	258.21	0.9657	0.04934	232.55	257.22	0.9530
15	0.05676	236.24	261.78	0.9782	0.05056	235.57	260.85	0.9657
20	0.05805	239.22	265.34	0.9904	0.05175	238.59	264.47	0.9781
25	0.05933	242.20	268.90	1.0025	0.05293	241.61	268.07	0.9903
30	0.06059	245.19	272.46	1.0143	0.05409	244.63	271.68	1.0023
35	0.06184	248.19	276.02	1.0259	0.05523	247.66	275.28	1.0141
40	0.06308	251.20	279.59	1.0374	0.05636	250.70	278.89	1.0257
45	0.06430	254.23	283.17	1.0488	0.05748	253.76	282.50	1.0371
50	0.06552	257.28	286.76	1.0600	0.05859	256.82	286.12	1.0484
55	0.06672	260.34	290.36	1.0710	0.05969	259.90	289.75	1.0595

T °C	$v \times 10^3$ m³/kg	u kJ/kg	h kJ/kg	s kJ/kg K	$v \times 10^3$ m³/kg	u kJ/kg	h kJ/kg	s kJ/kg K
	p = 5.5 bar = 0.55 MPa (T_{sat} = 3.08°C)				p = 6.0 bar = 0.60 MPa (T_{sat} = 5.85°C)			
Sat.	0.04271	227.53	251.02	0.9226	0.03923	228.44	251.98	0.9186
5	0.04317	228.72	252.46	0.9278	0.04015	231.05	255.14	0.9299
10	0.04433	231.81	256.20	0.9411	0.04122	234.18	258.91	0.9431
15	0.04547	234.89	259.90	0.9540	0.04227	237.29	262.65	0.9560
20	0.04658	237.95	263.57	0.9667	0.04330	240.39	266.37	0.9685
25	0.04768	241.01	267.23	0.9790	0.04431	243.49	270.07	0.9808
30	0.04875	244.07	270.88	0.9912	0.04530	246.58	273.76	0.9929
35	0.04982	247.13	274.53	1.0031	0.04628	249.68	277.45	1.0048
40	0.05086	250.20	278.17	1.0148	0.04724	252.78	281.13	1.0164

TABLE A-9 (Continued)

T °C	*v* × 10^3 m^3/kg	*u* kJ/kg	*h* kJ/kg	*s* kJ/kg K	*v* × 10^3 m^3/kg	*u* kJ/kg	*h* kJ/kg	*s* kJ/kg K
45	0.05190	253.27	281.82	1.0264	0.04820	255.90	284.82	1.0279
50	0.05293	256.36	285.47	1.0378	0.04914	259.02	288.51	1.0393
55	0.05394	259.46	289.13	1.0490	0.05008	262.15	292.20	1.0504
60	0.05495	262.58	292.80	1.0601				
	p = 7.0 bar 0.70 MPa (T_{sat} = 10.91°C)				p = 8.0 bar 0.80 MPa (T_{sat} = 15.45°C)			
Sat.	0.03371	230.04	253.64	0.9117	0.02953	231.43	255.05	0.9056
15	0.03451	232.70	256.86	0.9229	0.03033	234.47	258.74	0.9182
20	0.03547	235.92	260.75	0.9363	0.03118	237.76	262.70	0.9315
25	0.03639	239.12	264.59	0.9493	0.03202	241.04	266.66	0.9448
30	0.03730	242.29	268.40	0.9619	0.03283	24428	270.54	0.9574
35	0.03819	245.46	272.19	0.9743	0.03363	247.52	274.42	0.9700
40	0.03906	248.62	275.96	0.9865	0.03440	250.74	278.26	0.9821
45	0.03992	251.78	279.72	0.9984	0.03517	253.96	282.10	0.9941
50	0.04076	254.94	283.48	1.0101	0.03592	257.18	285.92	1.0058
55	0.04160	258.11	287.23	1.0216	0.03667	260.40	289.74	1.0174
60	0.04242	261.29	290.99	1.0330	0.03741	263.64	293.56	1.0287
65	0.04324	264.48	294.75	1.0442	0.03814	266.87	297.38	1.0400
70	0.04405	267.68	298.51	1.0552				
	p = 9.0 bar = 0.90 MPa (T_{sat} = 19.59°C)				p = 10.0 bar = 1.00 MPa (T_{sat} = 23.40°C)			
Sat.	0.02623	232.64	256.25	0.9001	0.02358	233.71	257.28	0.8952
20	0.02630	232.92	256.59	0.9013	0.02457	238.34	262.91	0.9139
30	0.02789	239.73	264.83	0.9289	0.02598	245.18	271.17	0.9407
40	0.02939	246.37	272.82	0.9549	0.02732	251.90	279.22	0.9660
50	0.03082	252.95	280.68	0.9795	0.02860	258.56	287.15	0.9902

TABLE A-9 (Continued)

T °C	$v \times 10^3$ m³/kg	u kJ/kg	h kJ/kg	s kJ/kg K	$v \times 10^3$ m³/kg	u kJ/kg	h kJ/kg	s kJ/kg K
60	0.03219	259.49	288.46	1.0033				
70	0.03353	266.04	296.21	1.0262	0.02984	265.19	295.03	1.0135
80	0.03483	272.62	303.96	1.0484	0.03104	271.84	302.88	1.0361
90	0.03611	279.23	311.73	1.0701	0.03221	278.52	310.74	1.0580
100	0.03736	285.90	319.53	1.0913	0.03337	285.24	318.61	1.0794
110	0.03860	292.63	327.37	1.1120	0.03450	292.02	326.52	1.1003
120	0.03982	299.42	335.26	1.1323	0.03562	298.85	334.46	1.1207
130	0.04103	306.28	343.21	1.1523	0.03672	305.74	342.46	1.1408
140	0.04223	313.21	351.22	1.1719	0.03781	312.70	350.51	1.1605
150	0.04342	320.21	359.29	1.1912	0.03889	319.74	358.63	1.1790

p = 12.0 bar = 1.20 MPa (T_{sat} = 30.25°C) | p = 14.0 bar = 1.40 MPa (T_{sat} = 36.29°C)

T °C	$v \times 10^3$ m³/kg	u kJ/kg	h kJ/kg	s kJ/kg K	$v \times 10^3$ m³/kg	u kJ/kg	h kJ/kg	s kJ/kg K
Sat.	0.01955	235.48	258.94	0.8864	0.01662	236.89	260.16	0.8786
40	0.02083	242.63	267.62	0.9146	0.01708	239.78	263.70	0.8900
50	0.02204	249.69	276.14	0.9413	0.01823	247.29	272.81	0.9186
60	0.02319	256.60	284.43	0.9666	0.01929	254.52	281.53	0.9452
70	0.02428	263.44	292.58	0.9907	0.02029	261.60	290.01	0.9703
80	0.02534	270.25	300.66	1.0139	0.02125	268.60	298.34	0.9942
90	0.02636	277.07	308.70	1.0363	0.02217	275.56	306.60	1.0172
100	0.02736	283.90	316.73	1.0582	0.02306	282.52	314.80	1.0395
110	0.02834	290.77	324.78	1.0794	0.02393	289.49	323.00	1.0612
120	0.02930	297.69	332.85	1.1002	0.02478	296.50	331.19	1.0823
130	0.03024	304.65	340.95	1.1205	0.02562	303.55	339.41	1.1029
140	0.03118	311.68	349.09	1.1405	0.02644	310.64	347.65	1.1231
150	0.03210	318.77	357.29	1.1601	0.02725	317.79	355.94	1.1429
160	0.03301	325.92	365.54	1.1793	0.02805	324.99	364.26	1.1624
170	0.03392	333.14	373.84	1.1983	0.02884	332.26	372.64	1.1815

TABLE A-9 (Continued)

T °C	$p = 16.0$ bar $= 1.60$ MPa ($T_{sat} = 41.73°C$)				$p = 18.0$ bar $= 1.80$ MPa ($T_{sat} = 46.69°C$)			
	$v \times 10^3$ m³/kg	u kJ/kg	h kJ/kg	s kJ/kg K	$v \times 10^3$ m³/kg	u kJ/kg	h kJ/kg	s kJ/kg K
Sat.	0.01440	238.00	261.04	0.8715	0.01265	238.86	261.64	0.8649
50	0.01533	244.66	269.18	0.8971	0.01301	241.72	265.14	0.8758
60	0.01634	252.29	278.43	0.9252	0.01401	249.86	275.09	0.9061
70	0.01728	259.65	287.30	0.9515	0.01492	257.57	284.43	0.9337
80	0.01817	266.86	295.93	0.9762	0.01576	265.04	293.40	0.9595
90	0.01901	274.00	304.42	0.9999	0.01655	272.37	302.16	0.9839
100	0.01983	281.09	312.82	1.0228	0.01731	279.62	310.77	1.0073
110	0.02062	288.18	321.17	1.0448	0.01804	286.83	319.30	1.0299
120	0.02139	295.28	329.51	1.0663	0.01874	294.04	327.78	1.0517
130	0.02214	302.41	337.84	1.0872	0.01943	301.26	336.24	1.0730
140	0.02288	309.58	346.19	1.1077	0.02011	308.50	344.70	1.0937
150	0.02361	316.79	354.56	1.1277	0.02077	315.78	353.17	1.1139
160	0.02432	324.05	362.97	1.1473	0.02142	323.10	361.66	1.1338
170	0.02503	331.37	371.42	1.1666	0.02207	330.47	370.19	1.1532

T °C	$p = 20.0$ bar $= 2.00$ MPa ($T_{sat} = 51.26°C$)				$p = 24.0$ bar $= 2.4$ MPa ($T_{sat} = 59.46°C$)			
	$v \times 10^3$ m³/kg	u kJ/kg	h kJ/kg	s kJ/kg K	$v \times 10^3$ m³/kg	u kJ/kg	h kJ/kg	s kJ/kg K
Sat.	0.01124	239.51	261.98	0.8586	0.00907	240.22	261.99	0.8463
60	0.01212	247.20	271.43	0.8873	0.00913	240.78	262.68	0.8484
70	0.01300	255.35	281.36	0.9167	0.01006	250.30	274.43	0.8831
80	0.01381	263.12	290.74	0.9436	0.01085	258.89	284.93	0.9133
90	0.01457	270.67	299.80	0.9689	0.01156	267.01	294.75	0.9407
100	0.01528	278.09	308.65	0.9929	0.01222	274.85	304.18	0.9663
110	0.01596	285.44	317.37	1.0160	0.01284	282.53	313.35	0.9906
120	0.01663	292.76	326.01	1.0383	0.01343	290.11	322.35	1.0137

TABLE A-9 (Continued)

T °C	$v \times 10^3$ m³/kg	u kJ/kg	h kJ/kg	s kJ/kg K	$v \times 10^3$ m³/kg	u kJ/kg	h kJ/kg	s kJ/kg K
130	0.01727	300.08	334.61	1.0598	0.01400	297.64	331.25	1.0361
140	0.01789	307.40	343.19	1.0808	0.01456	305.14	340.08	1.0577
150	0.01850	314.75	351.76	1.1013	0.01509	312.64	348.87	1.0787
160	0.01910	322.14	360.34	1.1214	0.01562	320.16	357.64	1.0992
170	0.01969	329.56	368.95	1.1410	0.01613	327.70	366.41	1.1192
180	0.02027	337.03	377.58	1.1603	0.01663	335.27	375.20	1.1388

TABLE A-10 Properties of Saturated Refrigerant 134a (Liquid-Vapor): Temperature Table

Temp. °C	Press bar	Specific volume m³/kg		Internal Energy kJ/kg		Enthalpy kJ/kg			Entropy kJ/kg K		Temp °C
		Sat. Liquid $v_f \times 10^3$	Sat. Vapor v_g	Sat. Liquid u_f	Sat. Vapor u_g	Sat. Liquid h_f	Evap. h_{fg}	Sat. Vapor h_g	Sat. Liquid s_f	Sat. Vapor s_g	
−40	0.5164	0.7055	0.3569	−0.04	204.45	0.00	222.88	222.88	0.0000	0.9560	−40
−36	0.6332	0.7113	0.2947	4.68	206.73	4.73	220.67	225.40	0.0201	0.9506	−36
−32	0.7704	0.7172	0.2451	9.47	209.01	9.52	218.37	227.90	0.0401	0.9456	−32
−28	0.9305	0.7233	0.2052	14.31	211.29	14.37	216.01	230.38	0.0600	0.9411	−28
−26	1.0199	0.7265	0.1882	16.75	212.43	16.82	214.80	231.62	0.0699	0.9390	−26
−24	1.1160	0.7296	0.1728	19.21	213.57	19.29	213.57	232.85	0.0798	0.9370	−24
−22	1.2192	0.7328	0.1590	21.68	214.70	21.77	212.32	234.08	0.0897	0.9351	−22
−20	1.3299	0.7361	0.1464	24.17	215.84	24.26	211.05	235.31	0.0996	0.9332	−20
−18	1.4483	0.7395	0.1350	26.67	216.97	26.77	209.76	236.53	0.1094	0.9315	−18
−16	1.5748	0.7428	0.1247	29.18	218.10	29.30	208.45	237.74	0.1192	0.9298	−16
−12	1.8540	0.7498	0.1068	34.25	220.36	34.39	205.77	240.15	0.1388	0.9267	−12

TABLE A-10 (Continued)

Temp. °C	Press bar	Specific volume m³/kg		Internal Energy kJ/kg		Enthalpy kJ/kg			Entropy kJ/kg K		Temp °C
		Sat. Liquid $v_f \times 10^3$	Sat. Vapor v_g	Sat. Liquid u_f	Sat. Vapor u_g	Sat. Liquid h_f	Evap. h_g	Sat. Vapor h_g	Sat. Liquid s_f	Sat. Vapor s_g	
−8	2.1704	0.7569	0.0919	39.38	222.60	39.54	203.00	242.54	0.1583	0.9239	−8
−4	2.5274	0.7644	0.0794	44.56	224.84	44.75	200.15	244.90	0.1777	0.9213	−4
0	2.9282	0.7721	0.0689	49.79	227.06	50.02	197.21	247.23	0.1970	0.9190	0
4	3.3765	0.7801	0.0600	55.08	229.27	55.35	194.19	249.53	0.2162	0.9169	4
8	3.8756	0.7884	0.0525	60.43	231.46	60.73	191.07	251.80	0.2354	0.9150	8
12	4.4294	0.7971	0.0460	65.83	233.63	66.18	187.85	254.03	0.2545	0.9132	12
16	5.0416	0.8062	0.0405	71.29	235.78	71.69	184.52	256.22	0.2735	0.9116	16
20	5.7160	0.8157	0.0358	76.80	237.91	77.26	181.09	258.36	0.2924	0.9102	20
24	6.4566	0.8257	0.0317	82.37	240.01	82.90	177.55	260.45	0.3113	0.9089	24
26	6.8530	0.8309	0.0298	85.18	241.05	85.75	175.73	261.48	0.3208	0.9082	26
28	7.2675	0.8362	0.0281	88.00	242.08	88.61	173.89	262.50	0.3302	0.9076	28
30	7.7006	0.8417	0.0265	90.84	243.10	91.49	172.00	263.50	0.3396	0.9070	30
32	8.1528	0.8473	0.0250	93.70	244.12	94.39	170.09	264.48	0.3490	0.9064	32
34	8.6247	0.8530	0.0236	96.58	245.12	97.31	168.14	265.45	0.3584	0.9058	34
36	9.1168	0.8590	0.0223	99.47	246.11	100.25	166.15	266.40	0.3678	0.9053	36
38	9.6298	0.8651	0.0210	102.38	247.09	103.21	164.12	267.33	0.3772	0.9047	38
40	10.164	0.8714	0.0199	105.30	248.06	106.19	162.05	268.24	0.3866	0.9041	40
42	10.720	0.8780	0.0188	108.25	249.02	109.19	159.94	269.14	0.3960	0.9035	42
44	11.299	0.8847	0.0177	111.22	249.96	112.22	157.79	270.01	0.4054	0.9030	44
48	12.526	0.8989	0.0159	117.22	251.79	118.35	153.33	271.68	0.4243	0.9017	48
52	13.851	0.9142	0.0142	123.31	253.55	124.58	148.66	273.24	0.4432	0.9004	52
56	15.278	0.9308	0.0127	129.51	255.23	130.93	143.75	274.68	0.4622	0.8990	56
60	16.813	0.9488	0.0114	135.82	256.81	137.42	138.57	275.99	0.4814	0.8973	60

TABLE A-10 (Continued)

Temp. °C	Press. bar	Specific volume m³/kg		Internal Energy kJ/kg		Enthalpy kJ/kg			Entropy kJ/kg K		Temp. °C
		Sat. Liquid $v_f \times 10^3$	Sat. Vapor v_g	Sat. Liquid u_f	Sat. Vapor u_g	Sat. Liquid h_f	Evap. h_g	Sat. Vapor h_g	Sat. Liquid s_f	Sat. Vapor s_g	
70	21.162	1.0027	0.0086	152.22	260.15	154.34	124.08	278.43	0.5302	0.8918	70
80	26.324	1.0766	0.0064	169.88	262.14	172.71	106.41	279.12	0.5814	0.8827	80
90	32.435	1.1949	0.0046	189.82	261.34	193.69	82.63	276.32	0.6380	0.8655	90
100	39.742	1.5443	0.0027	218.60	248.49	224.74	34.40	259.13	0.7196	0.8117	100

TABLE A-11 Properties of Saturated Refrigerant 134a (Liquid-Vapor): Pressure Table

Press. bar	Temp. °C	Specific volume m³/kg		Internal Energy kJ/kg		Enthalpy kJ/kg			Entropy kJ/kg K		Press. bar
		Sat. Liquid $v_f \times 10^3$	Sat. Vapor v_g	Sat. Liquid u_f	Sat. Vapor u_g	Sat. Liquid h_f	Evap. h_g	Sat. Vapor h_g	Sat. Liquid s_f	Sat. Vapor s_g	
0.6	-37.07	0.7097	0.3100	3.41	206.12	3.46	221.27	224.72	0.0147	0.9520	0.6
0.8	-31.21	0.7184	0.2366	10.41	209.46	10.47	217.92	228.39	0.0440	0.9447	0.8
1.0	-26.43	0.7258	0.1917	16.22	212.18	16.29	215.06	231.35	0.0678	0.9395	1.0
1.2	-22.36	0.7323	0.1614	21.23	214.50	21.32	212.54	233.86	0.0879	0.9354	1.2
1.4	-18.80	0.7381	0.1395	25.66	216.52	25.77	210.27	236.04	0.1055	0.9322	1.4
1.6	-15.62	0.7435	0.1229	29.66	218.32	29.78	208.19	237.97	0.1211	0.9295	1.6
1.8	-12.73	0.7485	0.1098	33.31	219.94	33.45	206.26	239.71	0.1352	0.9273	1.8
2.0	-10.09	0.7532	0.0993	36.69	221.43	36.84	204.46	241.30	0.1481	0.9253	2.0
2.4	-5.37	0.7618	0.0834	42.77	224.07	42.95	201.14	244.09	0.1710	0.9222	2.4
2.8	-1.23	0.7697	0.0719	48.18	226.38	48.39	198.13	246.52	0.1911	0.9197	2.8

TABLE A-11 (Continued)

Press bar	Temp. °C	Specific volume m³/kg Sat. Liquid $v_f \times 10^3$	Sat. Vapor v_g	Internal Energy kJ/kg Sat. Liquid u_f	Sat. Vapor u_g	Enthalpy kJ/kg Sat. Liquid h_f	Evap. h_g	Sat. Vapor h_g	Entropy kJ/kg K Sat. Liquid s_f	Sat. Vapor s_g	Press bar
3.2	2.48	0.7770	0.0632	53.06	228.43	53.31	195.35	248.66	0.2089	0.9177	3.2
3.6	5.84	0.7839	0.0564	57.54	230.28	57.82	192.76	250.58	0.2251	0.9160	3.6
4.0	8.93	0.7904	0.0509	61.69	231.97	62.00	190.32	252.32	0.2399	0.9145	4.0
5.0	15.74	0.8056	0.0409	70.93	235.64	71.33	184.74	256.07	0.2723	0.9117	5.0
6.0	21.58	0.8196	0.0341	78.99	238.74	79.48	179.71	259.19	0.2999	0.9097	6.0
7.0	26.72	0.8328	0.0292	86.19	241.42	86.78	175.07	261.85	0.3242	0.9080	7.0
8.0	31.33	0.8454	0.0255	92.75	243.78	93.42	170.73	264.15	0.3459	0.9066	8.0
9.0	35.53	0.8576	0.0226	98.79	245.88	99.56	166.62	266.18	0.3656	0.9054	9.0
10.0	39.39	0.8695	0.0202	104.42	247.77	105.29	162.68	267.97	0.3838	0.9043	10.0
12.0	46.32	0.8928	0.0166	114.69	251.03	115.76	15523	270.99	0.4164	0.9023	12.0
14.0	52.43	0.9159	0.0140	123.98	253.74	125.26	148.14	273.40	0.4453	0.9003	14.0
16.0	57.92	0.9392	0.0121	132.52	256.00	134.02	141.31	275.33	0.4714	0.8982	16.0
18.0	62.91	0.9631	0.0105	140.49	257.88	142.22	134.60	276.83	0.4954	0.8959	18.0
20.0	67.49	0.9878	0.0093	148.02	259.41	149.99	127.95	277.94	0.5178	0.8934	20.0
25.0	77.59	1.0562	0.0069	165.48	261.84	168.12	111.06	279.17	0.5687	0.8854	25.0
30.0	86.22	1.1416	0.0053	181.88	262.16	185.30	92.71	278.01	0.6156	0.8735	30.0

TABLE A-12 Properties of Superheated Refrigerant 134a Vapor

p = 0.6 bar = 0.06 MPa (T_sat = −37.07°C)

T °C	v m³/kg	u kJ/kg	h kJ/kg	s kJ/kg K
Sat.	0.31003	206.12	224.72	0.9520
−20	0.33536	217.86	237.98	1.0062
−10	0.34992	224.97	245.96	1.0371
0	0.36433	232.24	254.10	1.0675
10	0.37861	239.69	262.41	1.0973
20	0.39279	247.32	270.89	1.1267
30	0.40688	255.12	279.53	1.1557
40	0.42091	263.10	288.35	1.1844
50	0.43487	271.25	297.34	1.2126
60	0.44879	279.58	306.51	1.2405
70	0.46266	288.08	315.84	1.2681
80	0.47650	296.75	325.34	1.2954
90	0.49031	305.58	335.00	1.3224

p = 1.0 bar = 0.10 MPa (T_sat = −26.43°C)

T °C	v m³/kg	u kJ/kg	h kJ/kg	s kJ/kg K
Sat.	0.19170	212.18	231.35	0.9395
−20	0.19770	216.77	236.54	0.9602
−10	0.20686	224.01	244.70	0.9918
0	0.21587	231.41	252.99	1.0227
10	0.22473	238.96	261.43	1.0531
20	0.23349	246.67	270.02	1.0829
30	0.24216	254.54	278.76	1.1122
40	0.25076	262.58	287.66	1.1411
50	0.25930	270.79	296.72	1.1696
60	0.26779	279.16	305.94	1.1977
70	0.27623	287.70	315.32	1.2254
80	0.28464	296.40	324.87	1.2528
90	0.29302	305.27	334.57	1.2799

p = 1.4 bar = 0.14 MPa (T_sat = −18.80°C)

T °C	v m³/kg	u kJ/kg	h kJ/kg	s kJ/kg K
Sat.	0.13945	216.52	236.04	0.9322
−10	0.14549	223.03	243.40	0.9606
0	0.15219	230.55	251.86	0.9922
10	0.15875	238.21	260.43	1.0230
20	0.16520	246.01	269.13	1.0532
30	0.17155	253.96	277.97	1.0828
40	0.17783	262.06	286.96	1.1120
50	0.18404	270.32	296.09	1.1407
60	0.19020	278.74	305.37	1.1690

p = 1.8 bar = 0.18 MPa (T_sat = −12.73°C)

T °C	v m³/kg	u kJ/kg	h kJ/kg	s kJ/kg K
Sat.	0.10983	219.94	239.71	0.9273
−10	0.11135	222.02	242.06	0.9362
0	0.11678	229.67	250.69	0.9684
10	0.12207	237.44	259.41	0.9998
20	0.12723	245.33	268.23	1.0304
30	0.13230	253.36	277.17	1.0604
40	0.13730	261.53	286.24	1.0898
50	0.14222	269.85	295.45	1.1187
60	0.14710	278.31	304.79	1.1472

TABLE A-12 (Continued)

T °C	v m³/kg	u kJ/kg	h kJ/kg	s kJ/kg·K
70	0.19633	287.32	314.80	1.1969
80	0.20241	296.06	324.39	1.2244
90	0.20846	304.95	334.14	1.2516
100	0.21449	314.01	344.04	1.2785

$p = 2.0$ bar $= 0.20$ MPa ($T_{sat} = -10.09°C$)

T °C	v m³/kg	u kJ/kg	h kJ/kg	s kJ/kg·K
Sat.	0.09933	221.43	241.30	0.9253
−10	0.09938	221.50	241.38	0.9256
0	0.10438	229.23	250.10	0.9582
10	0.10922	237.05	258.89	0.9898
20	0.11394	244.99	267.78	1.0206
30	0.11856	253.06	276.77	1.0508
40	0.12311	261.26	285.88	1.0804
50	0.12758	269.61	295.12	1.1094
60	0.13201	278.10	304.50	1.1380
70	0.13639	286.74	314.02	1.1661
80	0.14073	295.53	323.68	1.1939
90	0.14504	304.47	333.48	1.2212
100	0.14932	313.57	343.43	1.2483

$p = 2.8$ bar $= 0.28$ MPa ($T_{sat} = 1.23°C$)

T °C	v m³/kg	u kJ/kg	h kJ/kg	s kJ/kg·K
Sat.	0.07193	226.38	246.52	0.9197
0	0.07240	227.37	247.64	0.9238
10	0.07613	235.44	256.76	0.9566
20	0.07972	243.59	265.91	0.9883
30	0.08320	251.83	275.12	1.0192

T °C	v m³/kg	u kJ/kg	h kJ/kg	s kJ/kg·K
70	0.15193	286.93	314.28	1.1753
80	0.15672	295.71	323.92	1.2030
90	0.16148	304.63	333.70	1.2303
100	0.16622	313.72	343.63	1.2573

$p = 2.4$ bar $= 0.24$ MPa ($T_{sat} = -5.37°C$)

T °C	v m³/kg	u kJ/kg	h kJ/kg	s kJ/kg·K
Sat.	0.08343	224.07	244.09	0.9222
0	0.08574	228.31	248.89	0.9399
10	0.08993	236.26	257.84	0.9721
20	0.09399	244.30	266.85	1.0034
30	0.09794	252.45	275.95	1.0339
40	0.10181	260.72	285.16	1.0637
50	0.10562	269.12	294.47	1.0930
60	0.10937	277.67	303.91	1.1218
70	0.11307	286.35	313.49	1.1501
80	0.11674	295.18	323.19	1.1780
90	0.12037	304.15	333.04	1.2055
100	0.12398	313.27	343.03	1.2326

$p = 3.2$ bar $= 0.32$ MPa ($T_{sat} = 2.48°C$)

T °C	v m³/kg	u kJ/kg	h kJ/kg	s kJ/kg·K
Sat.	0.06322	228.43	248.66	0.9177
0	0.06576	234.61	255.65	0.9427
10	0.06901	242.87	264.95	0.9749
20	0.07214	251.19	274.28	1.0062
30	0.07518	259.61	283.67	1.0367

TABLE A-12 (Continued)

T °C	v m³/kg	u kJ/kg	h kJ/kg	s kJ/kg K		v m³/kg	u kJ/kg	h kJ/kg	s kJ/kg K
40	0.08660	260.17	284.42	1.0494		0.07815	268.14	293.15	1.0665
50	0.08992	268.64	293.81	1.0789		0.08106	276.79	302.72	1.0957
60	0.09319	277.23	303.32	1.1079		0.08392	285.56	312.41	1.1243
70	0.09641	285.96	312.95	1.1364		0.08674	294.46	322.22	1.1525
80	0.09960	294.82	322.71	1.1644		0.08953	303.50	332.15	1.1802
90	0.10275	303.83	332.60	1.1920		0.09229	312.68	342.21	1.2076
100	0.10587	312.98	342.62	1.2193		0.09503	322.00	352.40	1.2345
110	0.10897	322.27	352.78	1.2461		0.09774	331.45	362.73	1.2611
120	0.11205	331.71	363.08	1.2727					

		p = 4.0 bar = 0.40 MPa (T_{sat} = 8.93°C)					p = 5.0 bar = 0.50 MPa (T_{sat} = 15.74°C)		
T °C	v m³/kg	u kJ/kg	h kJ/kg	s kJ/kg K		v m³/kg	u kJ/kg	h kJ/kg	s kJ/kg K
Sat.	0.05089	231.97	252.32	0.9145		0.04086	235.64	256.07	0.9117
10	0.05119	232.87	253.35	0.9182		0.04188	239.40	260.34	0.9264
20	0.05397	241.37	262.96	0.9515		0.04416	248.20	270.28	0.9597
30	0.05662	249.89	272.54	0.9837		0.04633	256.99	280.16	0.9918
40	0.05917	258.47	282.14	1.0148		0.04842	265.83	290.04	1.0229
50	0.06164	267.13	291.79	1.0452		0.05043	274.73	299.95	1.0531
60	0.06405	275.89	301.51	1.0748		0.05240	283.72	309.92	1.0825
70	0.06641	284.75	311.32	1.1038		0.05432	292.80	319.96	1.1114
80	0.06873	293.73	321.23	1.1322		0.05620	302.00	330.10	1.1397
90	0.07102	302.84	331.25	1.1602		0.05805	311.31	340.33	1.1675
100	0.07327	312.07	341.38	1.1878		0.05988	320.74	350.68	1.1949
110	0.07550	321.44	351.64	1.2149		0.06168	330.30	361.14	1.2218
120	0.07771	330.94	362.03	1.2417		0.06347	339.98	371.72	1.2484
130	0.07991	340.58	372.54	1.2681		0.06524	349.79	382.42	1.2746
140	0.08208	350.35	383.18	1.2941					

TABLE A-12 (Continued)

T °C	v m³/kg	u kJ/kg	h kJ/kg	s kJ/kg K	v m³/kg	u kJ/kg	h kJ/kg	s kJ/kg K
	p = 6.0 bar = 0.60 MPa (Tsat = 21.58°C)				p = 7.0 bar = 0.70 MPa (Tsat = 26.72°C)			
Sat.	0.03408	238.74	259.19	0.9097	0.02918	241.42	261.85	0.9080
30	0.03581	246.41	267.89	0.9388	0.02979	244.51	265.37	0.9197
40	0.03774	255.45	278.09	0.9719	0.03157	253.83	275.93	0.9539
50	0.03958	264.48	288.23	1.0037	0.03324	263.08	286.35	0.9867
60	0.04134	273.54	298.35	1.0346	0.03482	272.31	296.69	1.0182
70	0.04304	282.66	308.48	1.0645	0.03634	281.57	307.01	1.0487
80	0.04469	291.86	318.67	1.0938	0.03781	290.88	317.35	1.0784
90	0.04631	301.14	328.93	1.1225	0.03924	300.27	327.74	1.1074
100	0.04790	310.53	339.27	1.1505	0.04064	309.74	338.19	1.1358
110	0.04946	320.03	349.70	1.1781	0.04201	319.31	348.71	1.1637
120	0.05099	329.64	360.24	1.2053	0.04335	328.98	359.33	1.1910
130	0.05251	339.38	370.88	1.2320	0.04468	338.76	370.04	1.2179
140	0.05402	349.23	381.64	1.2584	0.04599	348.66	380.86	1.2444
150	0.05550	359.21	392.52	1.2844	0.04729	358.68	391.79	1.2706
160	0.05698	369.32	403.51	1.3100	0.04857	368.82	402.82	1.2963
	p = 8.0 bar = 0.80 MPa (Tsat = 31.33°C)				p = 9.0 bar = 0.90 MPa (Tsat = 35.53°C)			
Sat.	0.02547	243.78	264.15	0.9066	0.02255	245.88	266.18	0.9054
40	0.02691	252.13	273.66	0.9374	0.02325	250.32	271.25	0.9217
50	0.02846	261.62	284.39	0.9711	0.02472	260.09	282.34	0.9566
60	0.02992	271.04	294.98	1.0034	0.02609	269.72	293.21	0.9897
70	0.03131	280.45	305.50	1.0345	0.02738	279.30	303.94	1.0214
80	0.03264	289.89	316.00	1.0647	0.02861	288.87	314.62	1.0521
90	0.03393	299.37	326.52	1.0940	0.02980	298.46	325.28	1.0819
100	0.03519	308.93	337.08	1.1227	0.03095	308.11	335.96	1.1109

TABLE A-12 (Continued)

T °C	v m³/kg	u kJ/kg	h kJ/kg	s kJ/kg K	v m³/kg	u kJ/kg	h kJ/kg	s kJ/kg K
110	0.03642	318.57	347.71	1.1508	0.03207	317.82	346.68	1.1392
120	0.03762	328.31	358.40	1.1784	0.03316	327.62	357.47	1.1670
130	0.03881	338.14	369.19	1.2055	0.03423	337.52	368.33	1.1943
140	0.03997	348.09	380.07	1.2321	0.03529	347.51	379.27	1.2211
150	0.04113	358.15	391.05	1.2584	0.03633	357.61	390.31	1.2475
160	0.04227	368.32	402.14	1.2843	0.03736	367.82	401.44	1.2735
170	0.04340	378.61	413.33	1.3098	0.03838	378.14	412.68	1.2992
180	0.04452	389.02	424.63	1.3351	0.03939	388.57	424.02	1.3245
	p = 10.0 bar = 1.00 MPa (T_sat = 39.39°C)				p = 12.0 bar = 1.20 MPa (T_sat = 46.32°C)			
Sat.	0.02020	247.77	267.97	0.9043	0.01663	251.03	270.99	0.9023
40	0.02029	248.39	268.68	0.9066	0.01712	254.98	275.52	0.9164
50	0.02171	258.48	280.19	0.9428	0.01835	265.42	287.44	0.9527
60	0.02301	268.35	291.36	0.9768	0.01947	275.59	298.96	0.9868
70	0.02423	278.11	302.34	1.0093	0.02051	285.62	310.24	1.0192
80	0.02538	287.82	313.20	1.0405	0.02150	295.59	321.39	1.0503
90	0.02649	297.53	324.01	1.0707	0.02244	305.54	332.47	1.0804
100	0.02755	307.27	334.82	1.1000	0.02335	315.50	343.52	1.1096
110	0.02858	317.06	345.65	1.1286	0.02423	325.51	354.58	1.1381
120	0.02959	326.93	356.52	1.1567	0.02508	335.58	365.68	1.1660
130	0.03058	336.88	367.46	1.1841	0.02592	345.73	376.83	1.1933
140	0.03154	346.92	378.46	1.2111	0.02674	355.95	388.04	1.2201
150	0.03250	357.06	389.56	1.2376	0.02754	366.27	399.33	1.2465
160	0.03344	367.31	400.74	1.2638	0.02834	376.69	410.70	1.2724
170	0.03436	377.66	412.02	1.2895	0.02912	387.21	422.16	1.2980
180	0.03528	388.12	423.40	1.3149				

TABLE A-12 (Continued)

T °C	v m³/kg	u kJ/kg	h kJ/kg	s kJ/kg·K	v m³/kg	u kJ/kg	h kJ/kg	s kJ/kg·K
	p = 14.0 bar = 1.40 MPa (T_{sat} = 52.43°C)				p = 16.0 bar = 1.60 MPa (T_{sat} = 57.92°C)			
Sat.	0.01405	253.74	273.40	0.9003	0.01208	256.00	275.33	0.8982
60	0.01495	262.17	283.10	0.9297	0.01233	258.48	278.20	0.9069
70	0.01603	272.87	295.31	0.9658	0.01340	269.89	291.33	0.9457
80	0.01701	283.29	307.10	0.9997	0.01435	280.78	303.74	0.9813
90	0.01792	293.55	318.63	1.0319	0.01521	291.39	315.72	1.0148
100	0.01878	303.73	330.02	1.0628	0.01601	301.84	327.46	1.0467
110	0.01960	313.88	341.32	1.0927	0.01677	312.20	339.04	1.0773
120	0.02039	324.05	352.59	1.1218	0.01750	322.53	350.53	1.1069
130	0.02115	334.25	363.86	1.1501	0.01820	332.87	361.99	1.1357
140	0.02189	344.50	375.15	1.1777	0.01887	343.24	373.44	1.1638
150	0.02262	354.82	386.49	1.2048	0.01953	353.66	384.91	1.1912
160	0.02333	365.22	397.89	1.2315	0.02017	364.15	396.43	1.2181
170	0.02403	375.71	409.36	1.2576	0.02080	374.71	407.99	1.2445
180	0.02472	386.29	420.90	1.2834	0.02142	385.35	419.62	1.2704
190	0.02541	396.96	432.53	1.3088	0.02203	396.08	431.33	1.2960
200	0.02608	407.73	444.24	1.3338	0.02263	406.90	443.11	1.3212

TABLE A-13 Properties of Saturated Ammonia (Liquid-Vapor): Temperature Table

Temp. °C	Press. bar	Specific volume m³/kg		Internal Energy kJ/kg		Enthalpy kJ/kg			Entropy kJ/kg K		Temp °C
		Sat. Liquid $v_f \times 10^3$	Sat. Vapor v_g	Sat. Liquid u_f	Sat. Vapor u_g	Sat. Liquid h_f	Evap. h_{fg}	Sat. Vapor h_g	Sat. Liquid s_f	Sat. Vapor s_g	
−50	0.4086	1.4245	2.6265	−43.94	1264.99	−43.88	1416.20	1372.32	−0.1922	6.1543	−50
−45	0.5453	1.4367	2.0060	−22.03	1271.19	−21.95	1402.52	1380.57	−0.0951	6.0523	−45
−40	0.7174	1.4493	1.5524	−0.10	1277.20	0.00	1388.56	1388.56	0.0000	5.9557	−40
−36	0.8850	1.4597	1.2757	17.47	1281.87	17.60	1377.17	1394.77	0.0747	5.8819	−36
−32	1.0832	1.4703	1.0561	35.09	1286.41	35.25	1365.55	1400.81	0.1484	5.8111	−32
−30	1.1950	1.4757	0.9634	43.93	1288.63	44.10	1359.65	1403.75	0.1849	5.7767	−30
−28	1.3159	1.4812	0.8803	52.78	1290.82	52.97	1353.68	1406.66	0.2212	5.7430	−28
−26	1.4465	1.4867	0.8056	61.65	1292.97	61.86	1347.65	1409.51	0.2572	5.7100	−26
−22	1.7390	1.4980	0.6780	79.46	1297.18	79.72	1335.36	1415.08	0.3287	5.6457	−22
−20	1.9019	1.5038	0.6233	88.40	1299.23	88.68	1329.10	1417.79	0.3642	5.6144	−20
−18	2.0769	1.5096	0.5739	97.36	1301.25	97.68	1322.77	1420.45	0.3994	5.5837	−18
−16	2.2644	1.5155	0.5291	106.36	1303.23	106.70	1316.35	1423.05	0.4346	5.5536	−16
−14	2.4652	1.5215	0.4885	115.37	1305.17	115.75	1309.86	1425.61	0.4695	5.5239	−14
−12	2.6798	1.5276	0.4516	124.42	1307.08	124.83	1303.28	1428.11	0.5043	5.4948	−12
−10	2.9089	1.5338	0.4180	133.50	1308.95	133.94	1296.61	1430.55	0.5389	5.4662	−10
−8	3.1532	1.5400	0.3874	142.60	1310.78	143.09	1289.86	1432.95	0.5734	5.4380	−8
−6	3.4134	1.5464	0.3595	151.74	1312.57	152.26	1283.02	1435.28	0.6077	5.4103	−6
−4	3.6901	1.5528	0.3340	160.88	1314.32	161.46	1276.10	1437.56	0.6418	5.3831	−4
−2	3.9842	1.5594	0.3106	170.07	1316.04	170.69	1269.08	1439.78	0.6759	5.3562	−2
0	4.2962	1.5660	0.2892	179.29	1317.71	179.96	1261.97	1441.94	0.7097	5.3298	0
2	4.6270	1.5727	0.2695	188.53	1319.34	189.26	1254.77	1444.03	0.7435	5.3038	2
4	4.9773	1.5796	0.2514	197.80	1320.92	198.59	1247.48	1446.07	0.7770	5.2781	4
6	5.3479	1.5866	0.2348	207.10	1322.47	207.95	1240.09	1448.04	0.8105	5.2529	6

TABLE A-13 (Continued)

Temp. °C	Press bar	Specific volume m³/kg		Internal Energy kJ/kg		Enthalpy kJ/kg			Entropy kJ/kg K		Temp °C
		Sat. Liquid $v_f \times 10^3$	Sat. Vapor v_g	Sat. Liquid u_f	Sat. Vapor u_g	Sat. Liquid h_f	Evap. h_g	Sat. Vapor h_g	Sat. Liquid s_f	Sat. Vapor s_g	
8	5.7395	1.5936	0.2195	216.42	1323.96	217.34	1232.61	1449.94	0.8438	5.2279	8
10	6.1529	1.6008	0.2054	225.77	1325.42	226.75	1225.03	1451.78	0.8769	5.2033	10
12	6.5890	1.6081	0.1923	235.14	1326.82	236.20	1217.35	1453.55	0.9099	5.1791	12
16	7.5324	1.6231	0.1691	253.95	1329.48	255.18	1201.70	1456.87	0.9755	5.1314	16
20	8.5762	1.6386	0.1492	272.86	1331.94	274.26	1185.64	1459.90	1.0404	5.0849	20
24	9.7274	1.6547	0.1320	291.84	1334.19	293.45	1169.16	1462.61	1.1048	5.0394	24
28	10.993	1.6714	0.1172	310.92	1336.20	312.75	1152.24	1465.00	1.1686	4.9948	28
32	12.380	1.6887	0.1043	330.07	1337.97	332.17	1134.87	1467.03	1.2319	4.9509	32
36	13.896	1.7068	0.0930	349.32	1339.47	351.69	1117.00	1468.70	1.2946	4.9078	36
40	15.549	1.7256	0.0831	368.67	1340.70	371.35	1098.62	1469.97	1.3569	4.8652	40
45	17.819	1.7503	0.0725	393.01	1341.81	396.13	1074.84	1470.96	1.4341	4.8125	45
50	20.331	1.7765	0.0634	417.56	1342.42	421.17	1050.09	1471.26	1.5109	4.7604	50

TABLE A–14 Properties of Saturated Ammonia (Liquid-Vapor): Pressure Table

Press. bar	Temp. °C	Specific volume m³/kg		Internal Energy kJ/kg		Enthalpy kJ/kg			Entropy kJ/kg K		Press. bar
		Sat. Liquid $v_f \times 10^3$	Sat. Vapor v_g	Sat. Liquid u_f	Sat. Vapor u_g	Sat. Liquid h_f	Evap. h_g	Sat. Vapor h_g	Sat. Liquid s_f	Sat. Vapor s_g	
0.40	−50.36	1.4236	2.6795	−45.52	1264.54	−45.46	1417.18	1371.72	−0.1992	6.1618	0.40
0.50	−46.53	1.4330	2.1752	−28.73	1269.31	−28.66	1406.73	1378.07	−0.1245	6.0829	0.50
0.60	−43.28	1.4410	1.8345	−14.51	1273.27	−14.42	1397.76	1383.34	−0.0622	6.0186	0.60
0.70	−40.46	1.4482	1.5884	−21.11	1276.66	−2.011	1389.85	1387.84	−0.0086	5.9643	0.70
0.80	−37.94	1.4546	1.4020	8.93	1279.61	9.04	1382.73	1391.78	0.0386	5.9174	0.80
0.90	−35.67	1.4605	1.2559	18.91	1282.24	19.04	1376.23	1395.27	0.0808	5.8760	0.90
1.00	−33.60	1.4660	1.1381	28.03	1284.61	28.18	1370.23	1398.41	0.1191	5.8391	1.00
1.25	−29.07	1.4782	0.9237	48.03	1289.65	48.22	1356.89	1405.11	0.2018	5.7610	1.25
1.50	−25.22	1.4889	0.7787	65.10	1293.80	65.32	1345.28	1410.61	0.2712	5.6973	1.50
1.75	−21.86	1.4984	0.6740	80.08	1297.33	80.35	1334.92	1415.27	0.3312	5.6435	1.75
2.00	−18.86	1.5071	0.5946	93.50	1300.39	93.80	1325.51	1419.31	0.3843	5.5969	2.00
2.25	−16.15	1.5151	0.5323	105.68	1303.08	106.03	1316.83	1422.86	0.4319	5.5558	2.25
2.50	−13.67	1.5225	0.4821	116.88	1305.49	117.26	1308.76	1426.03	0.4753	5.5190	2.50
2.75	−11.37	1.5295	0.4408	127.26	1307.67	127.68	1301.20	1428.88	0.5152	5.4858	2.75
3.00	−9.24	1.5361	0.4061	136.96	1309.65	137.42	1294.05	1431.47	0.5520	5.4554	3.00
3.25	−7.24	1.5424	0.3765	146.06	1311.46	146.57	1287.27	1433.84	0.5864	5.4275	3.25
3.50	−5.36	1.5484	0.3511	154.66	1313.14	155.20	1280.81	1436.01	0.6186	5.4016	3.50
3.75	−3.58	1.5542	0.3289	162.80	1314.68	163.38	1274.64	1438.03	0.6489	5.3774	3.75
4.00	−1.90	1.5597	0.3094	170.55	1316.12	171.18	1268.71	1439.89	0.6776	5.3548	4.00
4.25	−0.29	1.5650	0.2921	177.96	1317.47	178.62	1263.01	1441.63	0.7048	5.3336	4.25
4.50	1.25	1.5702	0.2767	185.04	1318.73	185.75	1257.50	1443.25	0.7308	5.3135	4.50
4.75	2.72	1.5752	0.2629	191.84	1319.91	192.59	1252.18	1444.77	0.7555	5.2946	4.75
5.00	4.13	1.5800	0.2503	198.39	1321.02	199.18	1247.02	1446.19	0.7791	5.2765	5.00

TABLE A-14 (Continued)

Press. bar	Temp. °C	Specific volume m³/kg		Internal Energy kJ/kg		Enthalpy kJ/kg			Entropy kJ/kg K		Press. bar
		Sat. Liquid $v_f \times 10^3$	Sat. Vapor v_g	Sat. Liquid u_f	Sat. Vapor u_g	Sat. Liquid h_f	Evap. h_g	Sat. Vapor h_g	Sat. Liquid s_f	Sat. Vapor s_g	
5.25	5.48	1.5847	0.2390	204.69	1322.07	205.52	1242.01	1447.53	0.8018	5.2594	5.25
5.50	6.79	1.5893	0.2286	210.78	1323.06	211.65	1237.15	1448.80	0.8236	5.2430	5.50
5.75	8.05	1.5938	0.2191	216.66	1324.00	217.58	1232.41	1449.99	0.8446	5.2273	5.75
6.00	9.27	1.5982	0.2104	222.37	1324.89	223.32	1227.79	1451.12	0.8649	5.2122	6.00
7.00	13.79	1.6148	0.1815	243.56	1328.04	244.69	1210.38	1455.07	0.9394	5.1576	7.00
8.00	17.84	1.6302	0.1596	262.64	1330.64	263.95	1194.36	1458.30	1.0054	5.1099	8.00
9.00	21.52	1.6446	0.1424	280.05	1332.82	281.53	1179.44	1460.97	1.0649	5.0675	9.00
10.00	24.89	1.6584	0.1285	296.10	1334.66	297.76	1165.42	1463.18	1.1191	5.0294	10.00
12.00	30.94	1.6841	0.1075	324.99	1337.52	327.01	1139.52	1466.53	1.2152	4.9625	12.00
14.00	36.26	1.7080	0.0923	350.58	1339.56	352.97	1115.82	1468.79	1.2987	4.9050	14.00
16.00	41.03	1.7306	0.0808	373.69	1340.97	376.46	1093.77	1470.23	1.3729	4.8542	16.00
18.00	45.38	1.7522	0.0717	394.85	1341.88	398.00	1073.01	1471.01	1.4399	4.8086	18.00
20.00	49.37	1.7731	0.0644	414.44	1342.37	417.99	1053.27	1471.26	1.5012	4.7670	20.00

TABLE A-15 Properties of Superheated Ammonia Vapor

T °C	$p = 0.04$ bar $= 0.04$ MPa ($T_{sat} = -50.36$°C)				$p = 0.6$ bar $= 0.06$ MPa ($T_{sat} = -43.28$°C)			
	v m³/kg	u kJ/kg	h kJ/kg	s kJ/kg·K	v m³/kg	u kJ/kg	h kJ/kg	s kJ/kg·K
Sat.	2.6795	1264.54	1371.72	6.1618	1.8345	1273.27	1383.34	6.0186
-50	2.6841	1265.11	1372.48	6.1652				
-45	2.7481	1273.05	1382.98	6.2118				
-40	2.8118	1281.01	1393.48	6.2573	1.8630	1278.62	1390.40	6.0490
-35	2.8753	1288.96	1403.98	6.3018	1.9061	1286.75	1401.12	6.0946
-30	2.9385	1296.93	1414.47	6.3455	1.9491	1294.88	1411.83	6.1390
-25	3.0015	1304.90	1424.96	6.3882	1.9918	1303.01	1422.52	6.1826
-20	3.0644	1312.88	1435.46	6.4300	2.0343	1311.13	1433.19	6.2251
-15	3.1271	1320.87	1445.95	6.4711	2.0766	1319.25	1443.85	6.2668
-10	3.1896	1328.87	1456.45	6.5114	2.1188	1327.37	1454.50	6.3077
-5	3.2520	1336.88	1466.95	6.5509	2.1609	1335.49	1465.14	6.3478
0	3.3142	1344.90	1477.47	6.5898	2.2028	1343.61	1475.78	6.3871
5	3.3764	1352.95	1488.00	6.6280	2.2446	1351.75	1486.43	6.4257

T °C	$p = 0.8$ bar $= 0.08$ MPa ($T_{sat} = -37.94$°C)				$p = 1.0$ bar $= 0.10$ MPa ($T_{sat} = -33.60$°C)			
	v m³/kg	u kJ/kg	h kJ/kg	s kJ/kg·K	v m³/kg	u kJ/kg	h kJ/kg	s kJ/kg·K
Sat.	1.4021	1279.61	1391.78	5.9174	1.1381	1284.61	1398.41	5.8391
-35	1.4215	1284.51	1398.23	5.9446	1.1573	1290.71	1406.44	5.8723
-30	1.4543	1292.81	1409.15	5.9900	1.1838	1299.15	1417.53	5.9175
-25	1.4868	1301.09	1420.04	6.0343	1.2101	1307.57	1428.58	5.9616
-20	1.5192	1309.36	1430.90	6.0777	1.2362	1315.96	1439.58	6.0046
-15	1.5514	1317.61	1441.72	6.1200	1.2621	1324.33	1450.54	6.0467
-10	1.5834	1325.85	1452.53	6.1615	1.2880	1332.67	1461.47	6.0878
-5	1.6153	1334.09	1463.31	6.2021	1.3136	1341.00	1472.37	6.1281
0	1.6471	1342.31	1474.08	6.2419	1.3392	1349.33	1483.25	6.1676

TABLE A-15 (Continued)

T °C	v m³/kg	u kJ/kg	h kJ/kg	s kJ/kg K	v m³/kg	u kJ/kg	h kJ/kg	s kJ/kg K
5	1.6788	1350.54	1484.84	6.2809				
10	1.7103	1358.77	1495.60	6.3192	1.3647	1357.64	1494.11	6.2063
15	1.7418	1367.01	1506.35	6.3568	1.3900	1365.95	1504.96	6.2442
20	1.7732	1375.25	1517.10	6.3939	1.4153	1374.27	1515.80	6.2816
	$p = 1.5$ bar = 0.15 MPa ($T_{sat} = -25.22°C$)				$p = 2.0$ bar = 0.20 MPa ($T_{sat} = -18.86°C$)			
Sat.	0.7787	1293.80	1410.61	5.6973	0.59460	1300.39	1419.31	5.5969
−25	0.7795	1294.20	1411.13	5.6994				
−20	0.7978	1303.00	1422.67	5.7454				
−15	0.8158	1311.75	1434.12	5.7902	0.60542	1307.43	1428.51	5.6328
−10	0.8336	1320.44	1445.49	5.8338	0.61926	1316.46	1440.31	5.6781
−5	0.8514	1329.08	1456.79	5.8764	0.63294	1325.41	1452.00	5.7221
0	0.8689	1337.68	1468.02	5.9179	0.64648	1334.29	1463.59	5.7649
5	0.8864	1346.25	1479.20	5.9585	0.65989	1343.11	1475.09	5.8066
10	0.9037	1354.78	1490.34	5.9981	0.67320	1351.87	1486.51	5.8473
15	0.9210	1363.29	1501.44	6.0370	0.68640	1360.59	1497.87	5.8871
20	0.9382	1371.79	1512.51	6.0751	0.69952	1369.28	1509.18	5.9260
25	0.9553	1380.28	1523.56	6.1125	0.71256	1377.93	1520.44	5.9641
30	0.9723	1388.76	1534.60	6.1492	0.72553	1386.56	1531.67	6.0014
	$p = 2.5$ bar = 0.25 MPa ($T_{sat} = -13.67°C$)				$p = 3.0$ bar = 0.30 MPa ($T_{sat} = -9.24°C$)			
Sat.	0.48213	1305.49	1426.03	5.5190	0.40607	1309.65	1431.47	5.4554
−10	0.49051	1312.37	1435.00	5.5534	0.41428	1317.80	1442.08	5.4953
−5	0.50180	1321.65	1447.10	5.5989	0.42382	1327.28	1454.43	5.5409
0	0.51293	1330.83	1459.06	5.6431	0.43323	1336.64	1466.61	5.5851
5	0.52393	1339.91	1470.89	5.6860	0.44251	1345.89	1478.65	5.6280

TABLE A-15 (Continued)

T °C	v m³/kg	u kJ/kg	h kJ/kg	s kJ/kg K	v m³/kg	u kJ/kg	h kJ/kg	s kJ/kg K
10	0.53482	1348.91	1482.61	5.7278	0.45169	1355.05	1490.56	5.6697
15	0.54560	1357.84	1494.25	5.7685	0.46078	1364.13	1502.36	5.7103
20	0.55630	1366.72	1505.80	5.8083	0.46978	1373.14	1514.07	5.7499
25	0.56691	1375.55	1517.28	5.8471	0.47870	1382.09	1525.70	5.7886
30	0.57745	1384.34	1528.70	5.8851	0.48756	1391.00	1537.26	5.8264
35	0.58793	1393.10	1540.08	5.9223	0.49637	1399.86	1548.77	5.8635
40	0.59835	1401.84	1551.42	5.9589	0.50512	1408.70	1560.24	5.8998
45	0.60872	1410.56	1562.74	5.9947				

	$p = 3.5$ bar $= 0.35$ MPa ($T_{sat} = -5.36°C$)				$p = 4.0$ bar $= 0.40$ MPa ($T_{sat} = -1.90°C$)			
T °C	v m³/kg	u kJ/kg	h kJ/kg	s kJ/kg K	v m³/kg	u kJ/kg	h kJ/kg	s kJ/kg K
Sat.	0.35108	1313.14	1436.01	5.4016	0.30942	1316.12	1439.89	1439.89
0	0.36011	1323.66	1449.70	5.4522	0.31227	1319.95	1444.86	1444.86
10	0.37654	1342.82	1474.61	5.5417	0.32701	1339.68	1470.49	1470.49
20	0.39251	1361.49	1498.87	5.6259	0.34129	1358.81	1495.33	1495.33
30	0.40814	1379.81	1522.66	5.7057	0.35520	1377.49	1519.57	1519.57
40	0.42350	1397.87	1546.09	5.7818	0.36884	1395.85	1543.38	1543.38
60	0.45363	1433.55	1592.32	5.9249	0.39550	1431.97	1590.17	1590.17
80	0.48320	1469.06	1638.18	6.0586	0.42160	1467.77	1636.41	1636.41
100	0.51240	1504.73	1684.07	6.1850	0.44733	1503.64	1682.58	1682.58
120	0.54136	1540.79	1730.26	6.3056	0.47280	1539.85	1728.97	1728.97
140	0.57013	1577.38	1776.92	6.4213	0.49808	1576.55	1775.79	1775.79
160	0.59876	1614.60	1824.16	6.5330	0.52323	1613.86	1823.16	1823.16
180	0.62728	1652.51	1872.06	6.6411	0.54827	1651.85	1871.16	1871.16
200	0.65572	1691.15	1920.65	6.7460	0.57322	1690.56	1919.85	1919.85

TABLE A-15 (Continued)

T °C	v m³/kg	u kJ/kg	h kJ/kg	S kJ/kg K		v m³/kg	u kJ/kg	h kJ/kg	s kJ/kg K
	p = 4.5 bar = 0.45 MPa (T_{sat} = 1.25°C)					p = 5.0 bar = 0.50 MPa (T_{sat} = 4.13°C)			
Sat.	0.27671	1318.73	1443.25	5.3135		0.25034	1321.02	1446.19	5.2765
10	0.28846	1336.48	1466.29	5.3962		0.25757	1333.22	1462.00	5.3330
20	0.30142	1356.09	1491.72	5.4845		0.26949	1353.32	1488.06	5.4234
30	0.31401	1375.15	1516.45	5.5674		0.28103	1372.76	1513.28	5.5080
40	0.32631	1393.80	1540.64	5.6460		0.29227	1391.74	1537.87	5.5878
60	0.35029	1430.37	1588.00	5.7926		0.31410	1428.76	1585.81	5.7362
80	0.37369	1466.47	1634.63	5.9285		0.33535	1465.16	1632.84	5.8733
100	0.39671	1502.55	1681.07	6.0564		0.35621	1501.46	1679.56	6.0020
120	0.41947	1538.91	1727.67	6.1781		0.37681	1537.97	1726.37	6.1242
140	0.44205	1575.73	1774.65	6.2946		0.39722	1574.90	1773.51	6.2412
160	0.46448	1613.13	1822.15	6.4069		0.41749	1612.40	1821.14	6.3537
180	0.48681	1651.20	1870.26	6.5155		0.43765	1650.54	1869.36	6.4626
200	0.50905	1689.97	1919.04	6.6208		0.45771	1689.38	1918.24	6.5681
	p = 5.5 bar = 0.55 MPa (T_{sat} = 6.79°C)					p = 6.0 bar = 0.60 MPa (T_{sat} = 9.27°C)			
Sat.	0.22861	1323.06	1448.80	5.2430		0.21038	1324.89	1451.12	5.2122
10	0.23227	1329.88	1457.63	5.2743		0.21115	1326.47	1453.16	5.2195
20	0.24335	1350.50	1484.34	5.3671		0.22155	1347.62	1480.55	5.3145
30	0.25403	1370.35	1510.07	5.4534		0.23152	1367.90	1506.81	5.4026
40	0.26441	1389.64	1535.07	5.5345		0.24118	1387.52	1532.23	5.4851
50	0.27454	1408.53	1559.53	5.6114		0.25059	1406.67	1557.03	5.5631
60	0.28449	1427.13	1583.60	5.6848		0.25981	1425.49	1581.38	5.6373
80	0.30398	1463.85	1631.04	5.8230		0.27783	1462.52	1629.22	5.7768
100	0.32307	1500.36	1678.05	5.9525		0.29546	1499.25	1676.52	5.9071

TABLE A-15 (Continued)

T °C	v m³/kg	u kJ/kg	h kJ/kg	S kJ/kg K	v m³/kg	u kJ/kg	h kJ/kg	s kJ/kg K
120	0.34190	1537.02	1725.07	6.0753	0.31281	1536.07	1723.76	6.0304
140	0.36054	1574.07	1772.37	6.1926	0.32997	1573.24	1771.22	6.1481
160	0.37903	1611.66	1820.13	6.3055	0.34699	1610.92	1819.12	6.2613
180	0.39742	1649.88	1868.46	6.4146	0.36390	1649.22	1867.56	6.3707
200	0.41571	1688.79	1917.43	6.5203	0.38071	1688.20	1916.63	6.4766
	p = 7.0 bar = 0.70 MPa (T_sat = 13.79°C)				p = 8.0 bar = 0.80 MPa (T_sat = 17.84°C)			
Sat.	0.18148	1328.04	1455.07	5.1576	0.15958	1330.64	1458.30	5.1099
20	0.18721	1341.72	1472.77	5.2186	0.16138	1335.59	1464.70	5.1318
30	0.19610	1362.88	1500.15	5.3104	0.16948	1357.71	1493.29	5.2277
40	0.20464	1383.20	1526.45	5.3958	0.17720	1378.77	1520.53	5.3161
50	0.21293	1402.90	1551.95	5.4760	0.18465	1399.05	1546.77	5.3986
60	0.22101	1422.16	1576.87	5.5519	0.19189	1418.77	1572.28	5.4763
80	0.23674	1459.85	1625.56	5.6939	0.20590	1457.14	1621.86	5.6209
100	0.25205	1497.02	1673.46	5.8258	0.21949	1494.77	1670.37	5.7545
120	0.26709	1534.16	1721.12	5.9502	0.23280	1532.24	1718.48	5.8801
140	0.28193	1571.57	1768.92	6.0688	0.24590	1569.89	1766.61	5.9995
160	0.29663	1609.44	1817.08	6.1826	0.25886	1607.96	1815.04	6.1140
180	0.31121	1647.90	1865.75	6.2925	0.27170	1646.57	1863.94	6.2243
200	0.32571	1687.02	1915.01	6.3988	0.28445	1685.83	1913.39	6.3311
	p = 9.0 bar = 0.90 MPa (T_sat = 21.52°C)				p = 10.0 bar = 1.00 MPa (T_sat = 24.89°C)			
Sat.	0.14239	1332.82	1460.97	5.0675	0.12852	1334.66	1463.18	5.0294
30	0.14872	1352.36	1486.20	5.1520	0.13206	1346.82	1478.88	5.0816
40	0.15582	1374.21	1514.45	5.2436	0.13868	1369.52	1508.20	5.1768
50	0.16263	1395.11	1541.47	5.3286	0.14499	1391.07	1536.06	5.2644

TABLE A-15 (Continued)

T °C	v m³/kg	u kJ/kg	h kJ/kg	s kJ/kg K		v m³/kg	u kJ/kg	h kJ/kg	s kJ/kg K
60	0.16922	1415.32	1567.61	5.4083		0.15106	1411.79	1562.86	5.3460
80	0.18191	1454.39	1618.11	5.5555		0.16270	1451.60	1614.31	5.4960
100	0.19416	1492.50	1667.24	5.6908		0.17389	1490.20	1664.10	5.6332
120	0.20612	1530.30	1715.81	5.8176		0.18478	1528.35	1713.13	5.7612
140	0.21788	1568.20	1764.29	5.9379		0.19545	1566.51	1761.96	5.8823
160	0.22948	1606.46	1813.00	6.0530		0.20598	1604.97	1810.94	5.9981
180	0.24097	1645.24	1862.12	6.1639		0.21638	1643.91	1860.29	6.1095
200	0.25237	1684.64	1911.77	6.2711		0.22670	1683.44	1910.14	6.2171
	$p = 12.0$ bar $= 1.20$ MPa ($T_{sat} = 30.94°C$)					$p = 14.0$ bar $= 1.40$ MPa ($T_{sat} = 36.26°C$)			
Sat.	0.10751	1337.52	1466.53	4.9625		0.09231	1339.56	1468.79	4.9050
40	0.11287	1359.73	1495.18	5.0553		0.09432	1349.29	1481.33	4.9453
60	0.12378	1404.54	1553.07	5.2347		0.10423	1396.97	1542.89	5.1360
80	0.13387	1445.91	1606.56	5.3906		0.11324	1440.06	1598.59	5.2984
100	0.14347	1485.55	1657.71	5.5315		0.12172	1480.79	1651.20	5.4433
120	0.15275	1524.41	1707.71	5.6620		0.12986	1520.41	1702.21	5.5765
140	0.16181	1563.09	1757.26	5.7850		0.13777	1559.63	1752.52	5.7013
160	0.17072	1601.95	1806.81	5.9021		0.14552	1598.92	1802.65	5.8198
180	0.17950	1641.23	1856.63	6.0145		0.15315	1638.53	1852.94	5.9333
200	0.18819	1681.05	1906.87	6.1230		0.16068	1678.64	1903.59	6.0427
220	0.19680	1721.50	1957.66	6.2282		0.16813	1719.35	1954.73	6.1485
240	0.20534	1762.63	2009.04	6.3303		0.17551	1760.72	2006.43	6.2513
260	0.21382	1804.48	2061.06	6.4297		0.18283	1802.78	2058.75	6.3513
280	0.22225	1847.04	2113.74	6.5267		0.19010	1845.55	2111.69	6.4488

TABLE A-15 (Continued)

T °C	v m³/kg	u kJ/kg	h kJ/kg	s kJ/kg K
	p = 16.0 bar = 1.60 MPa (T_{sat} = 41.03°C)			
Sat.	0.08079	1340.97	1470.23	4.8542
60	0.08951	1389.06	1532.28	5.0461
80	0.09774	1434.02	1590.40	5.2156
100	0.10539	1475.93	1644.56	5.3648
120	0.11268	1516.34	1696.64	5.5008
140	0.11974	1556.14	1747.72	5.6276
160	0.12663	1595.85	1798.45	5.7475
180	0.13339	1635.81	1849.23	5.8621
200	0.14005	1676.21	1900.29	5.9723
220	0.14663	1717.18	1951.79	6.0789
240	0.15314	1758.79	2003.81	6.1823
260	0.15959	1801.07	2056.42	6.2829
280	0.16599	1844.05	2109.64	6.3809
	p = 20.0 bar = 2.00 MPa (T_{sat} = 49.37°C)			
Sat.	0.06445	1342.37	1471.26	4.7670
60	0.06875	1372.05	1509.54	4.8838
80	0.07596	1421.36	1573.27	5.0696
100	0.08248	1465.89	1630.86	5.2283
120	0.08861	1508.03	1685.24	5.3703
140	0.09447	1549.03	1737.98	5.5012
160	0.10016	1589.65	1789.97	5.6241
180	0.10571	1630.32	1841.74	5.7409
200	0.11116	1671.33	1893.64	5.8530

v m³/kg	u kJ/kg	h kJ/kg	s kJ/kg K
p = 18.0 bar = 1.80 MPa (T_{sat} = 45.38°C)			
0.07174	1341.88	1471.01	4.8086
0.07801	1380.77	1521.19	4.9627
0.08565	1427.79	1581.97	5.1399
0.09267	1470.97	1637.78	5.2937
0.09931	1512.22	1690.98	5.4326
0.10570	1552.61	1742.88	5.5614
0.11192	1592.76	1794.23	5.6828
0.11801	1633.08	1845.50	5.7985
0.12400	1673.78	1896.98	5.9096
0.12991	1715.00	1948.83	6.0170
0.13574	1756.85	2001.18	6.1210
0.14152	1799.35	2054.08	6.2222
0.14724	1842.55	2107.58	6.3207

TABLE A-15 (Continued)

T °C	v m³/kg	u kJ/kg	h kJ/kg	s kJ/kg K
220	0.11652	1712.82	1945.87	5.9611
240	0.12182	1754.90	1998.54	6.0658
260	0.12706	1797.63	2051.74	6.1675
280	0.13224	1841.03	2105.50	6.2665

TABLE A-16 Properties of Saturated Propane (Liquid-Vapor): Temperature Table

Temp. °C	Press. bar	Specific volume m³/kg Sat. Liquid $v_f \times 10^3$	Sat. Vapor v_g	Internal Energy kJ/kg Sat. Liquid u_f	Sat. Vapor u_g	Enthalpy kJ/kg Sat. Liquid h_f	Evap. h_g	Sat. Vapor h_g	Entropy kJ/kg K Sat. Liquid s_f	Sat. Vapor s_g	Temp °C
-100	0.02888	1.553	11.27	-128.4	319.5	-128.4	480.4	352.0	-0.634	2.140	-100
-90	0.06426	1.578	5.345	-107.8	329.3	-107.8	471.4	363.6	-0.519	2.055	-90
-80	0.1301	1.605	2.774	-87.0	339.3	-87.0	462.4	375.4	-0.408	1.986	-80
-70	0.2434	1.633	1.551	-65.8	349.5	-65.8	453.1	387.3	-0.301	1.929	-70
-60	0.4261	1.663	0.9234	-44.4	359.9	-44.3	443.5	399.2	-0.198	1.883	-60
-50	0.7046	1.694	0.5793	-22.5	370.4	-22.4	433.6	411.2	-0.098	1.845	-50
-40	1.110	1.728	0.3798	-0.2	381.0	0.0	423.2	423.2	0.000	1.815	-40
-30	1.677	1.763	0.2585	22.6	391.6	22.9	412.1	435.0	0.096	1.791	-30
-20	2.444	1.802	0.1815	45.9	402.4	46.3	400.5	446.8	0.190	1.772	-20
-10	3.451	1.844	0.1309	69.8	413.2	70.4	388.0	458.4	0.282	1.757	-10
0	4.743	1.890	0.09653	94.2	423.8	95.1	374.5	469.6	0.374	1.745	0
4	5.349	1.910	0.08591	104.2	428.1	105.3	368.8	474.1	0.410	1.741	4
8	6.011	1.931	0.07666	114.3	432.3	115.5	362.9	478.4	0.446	1.737	8
12	6.732	1.952	0.06858	124.6	436.5	125.9	356.8	482.7	0.482	1.734	12

TABLE A-16 (Continued)

Temp. °C	Press. bar	Specific volume m³/kg		Internal Energy kJ/kg		Enthalpy kJ/kg			Entropy kJ/kg K		Temp °C
		Sat. Liquid $v_f \times 10^3$	Sat. Vapor v_g	Sat. Liquid u_f	Sat. Vapor u_g	Sat. Liquid h_f	Evap. h_g	Sat. Vapor h_g	Sat. Liquid s_f	Sat. Vapor s_g	
16	7.515	1.975	0.06149	135.0	440.7	136.4	350.5	486.9	0.519	1.731	16
20	8.362	1.999	0.05525	145.4	444.8	147.1	343.9	491.0	0.555	1.728	20
24	9.278	2.024	0.04973	156.1	448.9	158.0	337.0	495.0	0.591	1.725	24
28	10.27	2.050	0.04483	166.9	452.9	169.0	329.9	498.9	0.627	1.722	28
32	11.33	2.078	0.04048	177.8	456.7	180.2	322.4	502.6	0.663	1.720	32
36	12.47	2.108	0.03659	188.9	460.6	191.6	314.6	506.2	0.699	1.717	36
40	13.69	2.140	0.03310	200.2	464.3	203.1	306.5	509.6	0.736	1.715	40
44	15.00	2.174	0.02997	211.7	467.9	214.9	298.0	512.9	0.772	1.712	44
48	16.40	2.211	0.02714	223.4	471.4	227.0	288.9	515.9	0.809	1.709	48
52	17.89	2.250	0.02459	235.3	474.6	239.3	279.3	518.6	0.846	1.705	52
56	19.47	2.293	0.02227	247.4	477.7	251.9	269.2	521.1	0.884	1.701	56
60	21.16	2.340	0.02015	259.8	480.6	264.8	258.4	523.2	0.921	1.697	60
65	23.42	2.406	0.01776	275.7	483.6	281.4	243.8	525.2	0.969	1.690	65
70	25.86	2.483	0.01560	292.3	486.1	298.7	227.7	526.4	1.018	1.682	70
75	28.49	2.573	0.01363	309.5	487.8	316.8	209.8	526.6	1.069	1.671	75
80	31.31	2.683	0.01182	327.6	488.2	336.0	189.2	525.2	1.122	1.657	80
85	34.36	2.827	0.01011	347.2	486.9	356.9	164.7	521.6	1.178	1.638	85
90	37.64	3.038	0.008415	369.4	482.2	380.8	133.1	513.9	1.242	1.608	90
95	41.19	3.488	0.006395	399.8	467.4	414.2	79.5	493.7	1.330	1.546	95
96.7	42.48	4.535	0.004535	434.9	434.9	454.2	0.0	457.2	1.437	1.437	96.7

TABLE A-17 Properties of Saturated Propane (Liquid-Vapor): Pressure Table

Press bar	Temp. °C	Specific volume m³/kg		Internal Energy kJ/kg		Enthalpy kJ/kg			Entropy kJ/kg K		Press bar
		Sat. Liquid $v_f \times 10^3$	Sat. Vapor v_g	Sat. Liquid u_f	Sat. Vapor u_g	Sat. Liquid h_f	Evap. h_g	Sat. Vapor h_g	Sat. Liquid s_f	Sat. Vapor s_g	
0.05	−93.28	1.570	6.752	−114.6	326.0	−114.6	474.4	359.8	−0.556	2.081	0.05
0.10	−83.87	1.594	3.542	−95.1	335.4	−95.1	465.9	370.8	−0.450	2.011	0.10
0.25	−69.55	1.634	1.513	−64.9	350.0	−64.9	452.7	387.8	−0.297	1.927	0.25
0.50	−56.93	1.672	0.7962	−37.7	363.1	−37.6	440.5	402.9	−0.167	1.871	0.50
0.75	−48.68	1.698	0.5467	−19.6	371.8	−19.5	432.3	412.8	−0.085	1.841	0.75
1.00	−42.38	1.719	0.4185	−5.6	378.5	−5.4	425.7	420.3	−0.023	1.822	1.00
2.00	−25.43	1.781	0.2192	33.1	396.6	33.5	406.9	440.4	0.139	1.782	2.00
3.00	−14.16	1.826	0.1496	59.8	408.7	60.3	393.3	453.6	0.244	1.762	3.00
4.00	−5.46	1.865	0.1137	80.8	418.0	81.5	382.0	463.5	0.324	1.751	4.00
5.00	1.74	1.899	0.09172	98.6	425.7	99.5	372.1	471.6	0.389	1.743	5.00
6.00	7.93	1.931	0.07680	114.2	432.2	115.3	363.0	478.3	0.446	1.737	6.00
7.00	13.41	1.960	0.06598	128.2	438.0	129.6	354.6	484.2	0.495	1.733	7.00
8.00	18.33	1.989	0.05776	141.0	443.1	142.6	346.7	489.3	0.540	1.729	8.00
9.00	22.82	2.016	0.05129	152.9	447.6	154.7	339.1	493.8	0.580	1.726	9.00
10.00	26.95	2.043	0.04606	164.0	451.8	166.1	331.8	497.9	0.618	1.723	10.00
11.00	30.80	2.070	0.04174	174.5	455.6	176.8	324.7	501.5	0.652	1.721	11.00
12.00	34.39	2.096	0.03810	184.4	459.1	187.0	317.8	504.8	0.685	1.718	12.00
13.00	37.77	2.122	0.03499	193.9	462.2	196.7	311.0	507.7	0.716	1.716	13.00
14.00	40.97	2.148	0.03231	203.0	465.2	206.0	304.4	510.4	0.745	1.714	14.00
15.00	44.01	2.174	0.02997	211.7	467.9	215.0	297.9	512.9	0.772	1.712	15.00
16.00	46.89	2.200	0.02790	220.1	470.4	223.6	291.4	515.0	0.799	1.710	16.00
17.00	49.65	2.227	0.02606	228.3	472.7	232.0	285.0	517.0	0.824	1.707	17.00
18.00	52.30	2.253	0.02441	236.2	474.9	240.2	278.6	518.8	0.849	1.705	18.00

TABLE A-17 (Continued)

Press bar	Temp. °C	Specific volume m³/kg Sat. Liquid $v_f \times 10^3$	Sat. Vapor v_g	Internal Energy kJ/kg Sat. Liquid u_f	Sat. Vapor u_g	Enthalpy kJ/kg Sat. Liquid h_f	Evap. h_g	Sat. Vapor h_g	Entropy kJ/kg K Sat. Liquid s_f	Sat. Vapor s_g	Press bar
19.00	54.83	2.280	0.02292	243.8	476.9	248.2	272.2	520.4	0.873	1.703	19.00
20.00	57.27	2.308	0.02157	251.3	478.7	255.9	265.9	521.8	0.896	1.700	20.00
22.00	61.90	2.364	0.01921	265.8	481.7	271.0	253.0	524.0	0.939	1.695	22.00
24.00	66.21	2.424	0.01721	279.7	484.3	285.5	240.1	525.6	0.981	1.688	24.00
26.00	70.27	2.487	0.01549	293.1	486.2	299.6	226.9	526.5	1.021	1.681	26.00
28.00	74.10	2.555	0.01398	306.2	487.5	313.4	213.2	526.6	1.060	1.673	28.00
30.00	77.72	2.630	0.01263	319.2	488.1	327.1	198.9	526.0	1.097	1.664	30.00
35.00	86.01	2.862	0.009771	351.4	486.3	361.4	159.1	520.5	1.190	1.633	35.00
40.00	93.38	3.279	0.007151	387.9	474.7	401.0	102.3	503.3	1.295	1.574	40.00
42.48	96.70	4.535	0.004535	434.9	434.9	454.2	100.0	454.2	1.437	1.437	42.48

TABLE A-18 Properties of Superheated Propane Vapor

T °C	$v \times 10^3$ m³/kg	u kJ/kg	h kJ/kg	s kJ/kg K	$v \times 10^3$ m³/kg	u kJ/kg	h kJ/kg	s kJ/kg K
	p = 0.05 bar = 0.005 MPa (T_sat = 93.28°C)				p = 0.1 bar = 0.01 MPa (T_sat = −83.87°C)			
Sat.	6.752	326.0	359.8	2.081	3.542	367.3	370.8	2.011
−90	6.877	329.4	363.8	2.103	3.617	339.5	375.7	2.037
−80	7.258	339.8	376.1	2.169	3.808	350.3	388.4	2.101
−70	7.639	350.6	388.8	2.233	3.999	361.5	401.5	2.164
−60	8.018	361.8	401.9	2.296	4.190	373.1	415.0	2.226

TABLE A-18 (Continued)

T °C	v × 10³ m³/kg	u kJ/kg	h kJ/kg	s kJ/kg K	v × 10³ m³/kg	U kJ/kg	h kJ/kg	s kJ/kg K
−50	8.397	373.3	415.3	2.357				
−40	8.776	385.1	429.0	2.418	4.380	385.0	428.8	2.286
−30	9.155	397.4	443.2	2.477	4.570	397.3	443.0	2.346
−20	9.533	410.1	457.8	2.536	4.760	410.0	457.6	2.405
−10	9.911	423.2	472.8	2.594	4.950	423.1	472.6	2.463
0	10.29	436.8	488.2	2.652	5.139	436.7	488.1	2.520
10	10.67	450.8	504.1	2.709	5.329	450.6	503.9	2.578
20	11.05	270.6	520.4	2.765	5.518	465.1	520.3	2.634

T °C	v × 10³ m³/kg	u kJ/kg	h kJ/kg	s kJ/kg K	v × 10³ m³/kg	U kJ/kg	h kJ/kg	s kJ/kg K
	p = 0.5 bar = 0.05 MPa (Tsat = −56.93°C)				p = 1.0 bar = 0.1 MPa (Tsat = −42.38°C)			
Sat.	0.796	363.1	402.9	1.871	0.4185	378.5	420.3	1.822
−50	0.824	371.3	412.5	1.914	0.4234	381.5	423.8	1.837
−40	0.863	383.4	426.6	1.976	0.4439	394.2	438.6	1.899
−30	0.903	396.0	441.1	2.037	0.4641	407.3	453.7	1.960
−20	0.942	408.8	455.9	2.096	0.4842	420.7	469.1	2.019
−10	0.981	422.1	471.1	2.155	0.5040	434.4	484.8	2.078
0	1.019	435.8	486.7	2.213	0.5238	448.6	501.0	2.136
10	1.058	449.8	502.7	2.271	0.5434	463.3	517.6	2.194
20	1.096	464.3	519.1	2.328	0.5629	478.2	534.5	2.251
30	1.135	479.2	535.9	2.384	0.5824	493.7	551.9	2.307
40	1.173	494.6	5532	2.440	0.6018	509.5	569.7	2.363
50	1.211	510.4	570.9	2.496	0.6211	525.8	587.9	2.419
60	1.249	526.7	589.1	2.551				

TABLE A-18 (Continued)

T °C	v × 10³ m³/kg	u kJ/kg	h kJ/kg	s kJ/kg K	v × 10³ m³/kg	U kJ/kg	h kJ/kg	s kJ/kg K
	p = 2.0 bar = 0.2 MPa (T_sat = −25.43°C)				p = 3.0 bar = 0.3 MPa (T_sat = −14.16°C)			
Sat.	0.2192	396.6	440.4	1.782	0.1496	408.7	453.6	1.762
−20	0.2251	404.0	449.0	1.816				
−10	0.2358	417.7	464.9	1.877	0.1527	414.7	460.5	1.789
0	0.2463	431.8	481.1	1.938	0.1602	429.0	477.1	1.851
10	0.2566	446.3	497.6	1.997	0.1674	443.8	494.0	1.912
20	0.2669	461.1	514.5	2.056	0.1746	458.8	511.2	1.971
30	0.2770	476.3	531.7	2.113	0.1816	474.2	528.7	2.030
40	0.2871	491.9	549.3	2.170	0.1885	490.1	546.6	2.088
50	0.2970	507.9	567.3	2.227	0.1954	506.2	564.8	2.145
60	0.3070	524.3	585.7	2.283	0.2022	522.7	583.4	2.202
70	0.3169	541.1	604.5	2.339	0.2090	539.6	602.3	2.258
80	0.3267	558.4	623.7	2.394	0.2157	557.0	621.7	2.314
90	0.3365	576.1	643.4	2.449	0.2223	574.8	641.5	2.369
	p = 4.0 bar = 0.4 MPa (T_sat = −5.46°C)				p = 5.0 bar = 0.5 MPa (T_sat = 1.74°C)			
Sat.	0.1137	418.0	463.5	1.751	0.09172	425.7	471.6	1.743
0	0.1169	426.1	472.9	1.786	0.09577	438.4	486.3	1.796
10	0.1227	441.2	490.3	1.848	0.1005	454.1	504.3	1.858
20	0.1283	456.6	507.9	1.909	0.1051	470.0	522.5	1.919
30	0.1338	472.2	525.7	1.969	0.1096	486.1	540.9	1.979
40	0.1392	488.1	543.8	2.027	0.1140	502.5	559.5	2.038
50	0.1445	504.4	562.2	2.085	0.1183	519.4	578.5	2.095

TABLE A-18 (Continued)

T °C	v × 10³ m³/kg	u kJ/kg	h kJ/kg	s kJ/kg K	v × 10³ m³/kg	u kJ/kg	h kJ/kg	s kJ/kg K
60	0.1498	521.1	581.0	2.143	0.1226	536.6	597.9	2.153
70	0.1550	538.1	600.1	2.199	0.1268	554.1	617.5	2.209
80	0.1601	555.7	619.7	2.255	0.1310	572.1	637.6	2.265
90	0.1652	573.5	639.6	2.311	0.1351	590.5	658.0	2.321
100	0.1703	591.8	659.9	2.366	0.1392	609.3	678.9	2.376
110	0.1754	610.4	680.6	2.421				

p = 6.0 bar = 0.6 MPa (T_{sat} = 7.93°C) | p = 7.0 bar = 0.7 MPa (T_{sat} = 13.41°C)

T °C	v × 10³ m³/kg	u kJ/kg	h kJ/kg	s kJ/kg K	v × 10³ m³/kg	u kJ/kg	h kJ/kg	s kJ/kg K
Sat.	0.07680	432.2	478.3	1.737	0.06598	438.0	484.2	1.733
10	0.07769	435.6	482.2	1.751	0.06847	448.8	496.7	1.776
20	0.08187	451.5	500.6	1.815	0.07210	465.2	515.7	1.840
30	0.08588	467.7	519.2	1.877	0.07558	481.9	534.8	1.901
40	0.08978	484.0	537.9	1.938	0.07896	498.7	554.0	1.962
50	0.09357	500.7	556.8	1.997	0.08225	515.9	573.5	2.021
60	0.09729	517.6	576.0	2.056	0.08547	533.4	593.2	2.079
70	0.1009	535.0	595.5	2.113	0.08863	551.2	613.2	2.137
80	0.1045	552.7	615.4	2.170	0.09175	569.4	633.6	2.194
90	0.1081	570.7	635.6	2.227	0.09482	587.9	654.3	2.250
100	0.1116	589.2	656.2	2.283	0.09786	606.8	675.3	2.306
110	0.1151	608.0	677.1	2.338	0.1009	626.2	696.8	2.361
120	0.1185	627.3	698.4	2.393				

TABLE A-18 (Continued)

T °C	$v \times 10^3$ m³/kg	u kJ/kg	h kJ/kg	s kJ/kg K	$v \times 10^3$ m³/kg	u kJ/kg	h kJ/kg	s kJ/kg K
	p = 8.0 bar = 0.8 MPa (T_{sat} = 18.33°C)				p = 9.0 bar = 0.9 MPa (T_{sat} = 22.82°C)			
Sat.	0.05776	443.1	489.3	1.729	0.05129	447.2	493.8	1.726
20	0.05834	445.9	492.6	1.740	0.05355	460.0	508.2	1.774
30	0.06170	462.7	512.1	1.806	0.05653	477.2	528.1	1.839
40	0.06489	479.6	531.5	1.869	0.05938	494.7	548.1	1.901
50	0.06796	496.7	551.1	1.930	0.06213	512.2	568.1	1.962
60	0.07094	514.0	570.8	1.990	0.06479	530.0	588.3	2.022
70	0.07385	531.6	590.7	2.049	0.06738	548.1	608.7	2.081
80	0.07669	549.6	611.0	2.107	0.06992	566.5	629.4	2.138
90	0.07948	567.9	631.5	2.165	0.07241	585.2	650.4	2.195
100	0.08222	586.5	652.3	2.221	0.07487	604.3	671.7	2.252
110	0.08493	605.6	673.5	2.277	0.07729	623.7	693.3	2.307
120	0.08761	625.0	695.1	2.333	0.07969	643.6	715.3	2.363
130	0.09026	644.8	717.0	2.388	0.08206	663.8	737.7	2.418
140	0.09289	665.0	739.3	2.442				
	p = 10.0 bar = 1.0 MPa (T_{sat} = 26.95°C)				p = 12.0 bar = 1.2 MPa (T_{sat} = 34.39°C)			
Sat.	0.04606	451.8	497.9	1.723	1.723	459.1	459.1	1.718
30	0.04696	457.1	504.1	1.744	1.744	469.4	469.4	1.757
40	0.04980	474.8	524.6	1.810	1.810	487.8	487.8	1.824
50	0.05248	492.4	544.9	1.874	1.874	506.1	506.1	1.889
60	0.05505	510.2	565.2	1.936	1.936	524.4	524.4	1.951
70	0.05752	528.2	585.7	1.997	1.997	543.1	543.1	2.012

TABLE A-18 (Continued)

T °C	v×10³ m³/kg	u kJ/kg	h kJ/kg	s kJ/kg K	v×10³ m³/kg	u kJ/kg	h kJ/kg	s kJ/kg K
80	0.05992	546.4	606.3	2.056	2.056	561.8	561.8	2.071
90	0.06226	564.9	627.2	2.114	2.114	580.9	580.9	2.129
100	0.06456	583.7	648.3	2.172	2.172	600.4	600.4	2.187
110	0.06681	603.0	669.8	2.228	2.228	620.1	620.1	2.244
120	0.06903	622.6	691.6	2.284	2.284	640.1	640.1	2.300
130	0.07122	642.5	713.7	2.340	2.340	660.6	660.6	2.355
140	0.07338	662.8	736.2	2.395	2.395			
	p = 14.0 bar = 1.4 MPa (T$_{sat}$ = 40.97°C)				p = 16.0 bar = 1.6 MPa (T$_{sat}$ = 46.89°C)			
Sat.	0.03231	465.2	510.4	1.714	0.02790	470.4	515.0	1.710
50	0.03446	482.6	530.8	1.778	0.02861	476.7	522.5	1.733
60	0.03664	501.6	552.9	1.845	0.03075	496.6	545.8	1.804
70	0.03869	520.4	574.6	1.909	0.03270	516.2	568.5	1.871
80	0.04063	539.4	596.3	1.972	0.03453	535.7	590.9	1.935
90	0.04249	558.6	618.1	2.033	0.03626	555.2	613.2	1.997
100	0.04429	577.9	639.9	2.092	0.03792	574.8	635.5	2.058
110	0.04604	597.5	662.0	2.150	0.03952	594.7	657.9	2.117
120	0.04774	617.5	684.3	2.208	0.04107	614.8	680.5	2.176
130	0.04942	637.7	706.9	2.265	0.04259	635.3	703.4	2.233
140	0.05106	658.3	729.8	2.321	0.04407	656.0	726.5	2.290
150	0.05268	679.2	753.0	2.376	0.04553	677.1	749.9	2.346
160	0.05428	700.5	776.5	2.431	0.04696	698.5	773.6	2.401

TABLE A-18 (Continued)

p = 18.0 bar = 1.8 MPa (T_{sat} = 52.30°C)

T °C	v × 10³ m³/kg	u kJ/kg	h kJ/kg	s kJ/kg K
Sat.	0.02441	474.9	518.8	1.705
60	0.02606	491.1	538.0	1.763
70	0.02798	511.4	561.8	1.834
80	0.02974	531.6	585.1	1.901
90	0.03138	551.5	608.0	1.965
100	0.03293	571.5	630.8	2.027
110	0.03443	591.7	653.7	2.087
120	0.03586	612.1	676.6	2.146
130	0.03726	632.7	699.8	2.204
140	0.03863	653.6	723.1	2.262
150	0.03996	674.8	746.7	2.318
160	0.04127	696.3	770.6	2.374
170	0.04256	718.2	794.8	2.429
180	0.04383	740.4	819.3	2.484

p = 22.0 bar = 2.2 MPa (T_{sat} = 61.90°C)

T °C	v × 10³ m³/kg	u kJ/kg	h kJ/kg	s kJ/kg K
Sat.	0.01921	481.8	524.0	1.695
70	0.02086	500.5	546.4	1.761
80	0.02261	522.4	572.1	1.834
90	0.02417	543.5	596.7	1.903
100	0.02561	564.5	620.8	1.969
110	0.02697	585.3	644.6	2.032

p = 20.0 bar = 2.0 MPa (T_{sat} = 57.27°C)

T °C	v × 10³ m³/kg	U kJ/kg	h kJ/kg	s kJ/kg K
Sat.	0.02157	478.7	521.8	1.700
60	0.02216	484.8	529.1	1.722
70	0.02412	506.3	554.5	1.797
80	0.02585	527.1	578.8	1.867
90	0.02744	547.6	602.5	1.933
100	0.02892	568.1	625.9	1.997
110	0.03033	588.5	649.2	2.059
120	0.03169	609.2	672.6	2.119
130	0.03299	630.0	696.0	2.178
140	0.03426	651.2	719.7	2.236
150	0.03550	672.5	743.5	2.293
160	0.03671	694.2	767.6	2.349
170	0.03790	716.2	792.0	2.404
180	0.03907	738.5	816.6	2.459

p = 24.0 bar = 2.4 MPa (T_{sat} = 66.21°C)

T °C	v × 10³ m³/kg	U kJ/kg	h kJ/kg	s kJ/kg K
Sat.	0.01721	484.3	525.6	1.688
70	0.01802	493.7	536.9	1.722
80	0.01984	517.0	564.6	1.801
90	0.02141	539.0	590.4	1.873
100	0.02283	560.6	615.4	1.941
110	0.02414	581.9	639.8	2.006

TABLE A-18 (Continued)

T °C	v × 10³ m³/kg	u kJ/kg	h kJ/kg	s kJ/kg K	v × 10³ m³/kg	U kJ/kg	h kJ/kg	s kJ/kg K
120	0.02826	606.2	668.4	2.093	0.02538	603.2	664.1	2.068
130	0.02949	627.3	692.2	2.153	0.02656	624.6	688.3	2.129
140	0.03069	648.6	716.1	2.211	0.02770	646.0	712.5	2.188
150	0.03185	670.1	740.2	2.269	0.02880	667.8	736.9	2.247
160	0.03298	691.9	764.5	2.326	0.02986	689.7	761.4	2.304
170	0.03409	714.1	789.1	2.382	0.03091	711.9	786.1	2.360
180	0.03517	736.5	813.9	2.437	0.03193	734.5	811.1	2.416
	p = 26.0 bar = 2.6 MPa (T_{sat} = 70.27°C)				p = 30.0 bar = 3.0 MPa (T_{sat} = 77.72°C)			
Sat.	0.01549	486.2	526.5	1.681	0.01263	488.2	526.0	1.664
80	0.01742	511.0	556.3	1.767	0.01318	495.4	534.9	1.689
90	0.01903	534.2	583.7	1.844	0.01506	522.8	568.0	1.782
100	0.02045	556.4	609.6	1.914	0.01654	547.2	596.8	1.860
110	0.02174	578.3	634.8	1.981	0.01783	570.4	623.9	1.932
120	0.02294	600.0	659.6	2.045	0.01899	593.0	650.0	1.999
130	0.02408	621.6	684.2	2.106	0.02007	615.4	675.6	2.063
140	0.02516	643.4	708.8	2.167	0.02109	637.7	701.0	2.126
150	0.02621	665.3	733.4	2.226	0.02206	660.1	726.3	2.186
160	0.02723	687.4	758.2	2.283	0.02300	682.6	751.6	2.245
170	0.02821	709.9	783.2	2.340	0.02390	705.4	777.1	2.303
180	0.02918	732.5	808.4	2.397	0.02478	728.3	802.6	2.360
190	0.03012	755.5	833.8	2.452	0.02563	751.5	828.4	2.417

TABLE A-18 (Continued)

T °C	$v \times 10^3$ m³/kg	u kJ/kg	h kJ/kg	s kJ/kg K	$v \times 10^3$ m³/kg	U kJ/kg	h kJ/kg	s kJ/kg K
	p = 35.0 bar = 3.5 MPa (T_{sat} = 86.01°C)				p = 40.0 bar = 4.0 MPa (T_{sat} = 93.38°C)			
Sat.	0.00977	486.3	520.5	1.633	0.00715	474.7	503.3	1.574
90	0.01086	502.4	540.5	1.688				
100	0.01270	532.9	577.3	1.788	0.00940	512.1	549.7	1.700
110	0.01408	558.9	608.2	1.870	0.01110	544.7	589.1	1.804
120	0.01526	583.4	636.8	1.944	0.01237	572.1	621.6	1.887
130	0.01631	607.0	664.1	2.012	0.01344	597.4	651.2	1.962
140	0.01728	630.2	690.7	2.077	0.01439	621.9	679.5	2.031
150	0.01819	653.3	717.0	2.140	0.01527	645.9	707.0	2.097
160	0.01906	676.4	743.1	2.201	0.01609	669.7	734.1	2.160
170	0.01989	699.6	769.2	2.261	0.01687	693.4	760.9	2.222
180	0.02068	722.9	795.3	2.319	0.01761	717.3	787.7	2.281
190	0.02146	746.5	821.6	2.376	0.01833	741.2	814.5	2.340
200	0.02221	770.3	848.0	2.433	0.01902	765.3	841.4	2.397

TABLE A-19 Properties of Selected Solids and Liquids: c_p and K

Substance	Specific Heat, c_p (kJ/kg K)	Density, ρ (kg/m^3)	Thermal Conductivity, K (W/m K)
Selected Solids, **300 K**			
Aluminum	0.903	2700	237
Coal, anthracite	1.260	1350	0.26
Copper	0.385	8930	401
Granite	0.775	2630	2.79
Iron	0.447	7870	80.2
Lead	0.129	1130	35.3
Sand	0.800	1520	0.27
Silver	0.235	1050	429
Soil	1.840	2050	0.52
Steel (AISI 302)	0.480	8060	15.1
Tin	0.227	7310	
Building Materials, **300 K**			66.6
Brick, common	0.835	1920	0.72
Concrete (stone mix)	0.880	2300	1.4
Glass, plate	0.750	2500	1.4
Hardboard, siding	1.170	640	0.094
Limestone	0.810	2320	2.15
Plywood	1.220	545	0.12
Softwoods (fir, pine)	1.380	510	0.12
Insulating Materials, **300 K**			
Blanket (glass fiber)	-	16	0.046
Cork	1.800	120	0.039
Duct liner (glass fiber, coated)	0.835	32	0.038
Polystyrene (extruded)	1.210	55	0.027
Vermiculite fill (flakes)	0.835	80	0.068
Saturated Liquids			
Ammonia, 300 K	4.818	599.8	0.465
Mercury, 300 K	0.139	1352.9	8.540
Refrigerant 22, 300 K	1.267	1183.1	0.085
Refrigerant 134a, 300 K	1.434	1199.7	0.081
Unused Engine Oil, 300 K	1.909	884.1	0.145
Water, 275 K	4.211	999.9	0.574
300 K	4.179	996.5	0.613
325 K	4.182	987.1	0.645
350 K	4.195	973.5	0.668
375 K	4.220	956.8	0.681
400 K	4.256	937.4	0.688

TABLE A-20 Ideal Gas Specific Heats of Some Common Gases

Temp. K	c_p	c_v	k	c_p	c_v	k	c_p	c_v	k	Temp. K
	Air			Nitrogen, N_2			Oxygen, O_2			
250	1.003	0.716	1.401	1.039	0.742	1.400	0.913	0.653	1.398	250
300	1.005	0.718	1.400	1.039	0.743	1.400	0.918	0.658	1.395	300
350	1.008	0.721	1.398	1.041	0.744	1.399	0.928	0.668	1.389	350
400	1.013	0.726	1.395	1.044	0.747	1.397	0.941	0.681	1.382	400
450	1.020	0.733	1.391	1.049	0.752	1.395	0.956	0.696	1.373	450
500	1.029	0.742	1.387	1.056	0.759	1.391	0.972	0.712	1.365	500
550	1.040	0.753	1.381	1.065	0.768	1.387	0.988	0.728	1.358	550
600	1.051	0.764	1.376	1.075	0.778	1.382	1.003	0.743	1.350	600
650	1.063	0.776	1.370	1.086	0.789	1.376	1.017	0.758	1.343	650
700	1.075	0.788	1.364	1.098	0.801	1.371	1.031	0.771	1.337	700
750	1.087	0.800	1.359	1.110	0.813	1.365	1.043	0.783	1.332	750
800	1.099	0.812	1.354	1.121	0.825	1.360	1.054	0.794	1.327	800
900	1.121	0.834	1.344	1.145	0.849	1.349	1.074	0.814	1.319	900
1000	1.142	0.855	1.336	1.167	0.870	1.341	1.090	0.830	1.313	1000

Temp. K	Carbon Dioxide, CO_2			Carbon Monoxide, CO			Hydrogen, H_2			Temp. K
250	0.791	0.602	1.314	1.039	0.743	1.400	14.051	9.927	1.416	250
300	0.846	0.657	1.288	1.040	0.744	1.399	14.307	10.183	1.405	300
350	0.895	0.706	1.268	1.043	0.746	1.398	14.427	10.302	1.400	350
400	0.939	0.750	1.252	1.047	0.751	1.395	14.476	10.352	1.398	400
450	0.978	0.790	1.239	1.054	0.757	1.392	14.501	10.377	1.398	450
500	1.014	0.825	1.229	1.063	0.767	1.387	14.513	10.389	1.397	500
550	1.046	0.857	1.220	1.075	0.778	1.382	14.530	10.405	1.396	550
600	1.075	0.886	1.213	1.087	0.790	1.376	14.546	10.422	1.396	600
650	1.102	0.913	1.207	1.100	0.803	1.370	14.571	10.447	1.395	650
700	1.126	0.937	1.202	1.113	0.816	1.364	14.604	10.480	1.394	700
750	1.148	0.959	1.197	1.126	0.829	1.358	14.645	10.521	1.392	750
800	1.169	0.980	1.193	1.139	0.842	1.353	14.695	10.570	1.390	800
900	1.204	1.015	1.186	1.163	0.866	1.343	14.822	10.698	1.385	900
1000	1.234	1.045	1.181	1.185	0.888	1.335	14.983	10.859	1.380	1000

TABLE A-21 Variation of cp with Temperature for Selected Ideal Gases

Gas	α	$\beta \times 10^3$	$\gamma \times 10^6$	$\delta \times 10^9$	$\varepsilon \times 10^{12}$
CO	3.710	−1.619	3.692	−2.032	0.240
CO_2	2.401	8.735	−6.607	2.002	0
H_2	3.057	2.677	−5.810	5.521	−1.812
H_2O	4.070	−1.108	4.152	−2.964	0.807
O_2	3.626	−1.878	7.055	−6.764	2.156
N_2	3.675	−1.208	2.324	−0.632	−0.226
Air	3.653	−1.337	3.294	−1.913	0.276
SO_2	3.267	5.324	0.684	−5.281	2.559
CH_4	3.826	−3.979	24.558	−22.733	6.963
C_2H_2	1.410	19.057	−24.501	16.391	−4.135
C_2H_4	1.426	11.383	7.989	−16.254	6.749
Monoatomic gases[a]	2.5	0	0	0	0

TABLE A-22 Ideal Gas Properties of Air

$T(K)$, h and u (kJ/kg), $s°$ (kJ/kg K)

T	h	u	s°	pr	vr
				when $\Delta s = 0$[1]	
200	199.97	142.56	1.29559	0.3363	1707.
210	209.97	149.69	1.34444	0.3987	1512.
220	219.97	156.82	1.39105	0.4690	1346.
230	230.02	164.00	1.43557	0.5477	1205.
240	240.02	171.13	1.47824	0.6355	1084.
250	250.05	178.28	1.51917	0.7329	979.
260	260.09	185.45	1.55848	0.8405	887.8
270	270.11	192.60	1.59634	0.9590	808.0
280	280.13	199.75	1.63279	1.0889	738.0
285	285.14	203.33	1.65055	1.1584	706.1
290	290.16	206.91	1.66802	1.2311	676.1
295	295.17	210.49	1.68515	1.3068	647.9
300	300.19	214.07	1.70203	1.3860	621.2
305	305.22	217.67	1.71865	1.4686	596.0
310	310.24	221.25	1.73498	1.5546	572.3
315	315.27	224.85	1.75106	1.6442	549.8
320	320.29	228.42	1.76690	1.7375	528.6
325	325.31	232.02	1.78249	1.8345	508.4
330	330.34	235.61	1.79783	1.9352	489.4
340	340.42	242.82	1.82790	2.149	454.1
350	350.49	250.02	1.85708	2.379	422.2
360	360.58	257.24	1.88543	2.626	393.4
370	370.67	264.46	1.91313	2.892	367.2
380	380.77	271.69	1.94001	3.176	343.4
390	390.88	278.93	1.96633	3.481	321.5
400	400.98	286.16	1.99194	3.806	301.6

T	h	u	s°	pr	vr
				when $\Delta s = 0$	
450	451.80	322.62	2.11161	5.775	223.6
460	462.02	329.97	2.13407	6.245	211.4
470	472.24	337.32	2.15604	6.742	200.1
480	482.49	344.70	2.17760	7.268	189.5
490	492.74	352.08	2.19876	7.824	179.7
500	503.02	359.49	2.21952	8.411	170.6
510	513.32	366.92	2.23993	9.031	162.1
520	523.63	374.36	2.25997	9.684	154.1
530	533.98	381.84	2.27967	10.37	146.7
540	544.	389.34	2.29906	11.10	139.7
550	554.74	396.86	2.31809	11.86	133.1
560	565.17	404.42	2.33685	12.66	127.0
570	575.59	411.97	2.35531	13.50	121.2
580	586.04	419.55	2.37348	14.38	115.7
590	596.52	427.15	2.39140	15.31	110.6
600	607.02	434.78	2.40902	16.28	105.8
610	617.53	442.42	2.42644	17.30	101.2
620	628.07	450.09	2.44356	18.36	96.92
630	638.63	457.78	2.46048	19.84	92.84
640	649.22	465.50	2.47716	20.64	88.99
650	659.84	473.25	2.49364	21.86	85.34
660	670.47	481.01	2.50985	23.13	81.89
670	681.14	488.81	2.52589	24.46	78.61
680	691.82	496.62	2.54175	25.85	75.50
690	702.52	504.45	2.55731	27.29	72.56
700	713.27	512.33	2.57277	28.80	69.76

TABLE A-22 (Continued)

$T(K)$, h and u (kJ/kg), $s°$ (kJ/kg K)

T	h	u	$s°$	when $\Delta s = 0^1$ p_r	v_r	T	h	u	$s°$	when $\Delta s = 0$ p_r	v_r
410	411.12	293.43	2.01699	4.153	283.3	710	724.04	520.23	2.58810	30.38	67.07
420	421.26	300.69	2.04142	4.522	266.6	720	734.82	528.14	2.60319	32.02	64.53
430	431.43	307.99	2.06533	4.915	251.1	730	745.62	536.07	2.61803	33.72	62.13
440	441.61	315.30	2.08870	5.332	236.8	740	756.40	544.02	2.63280	35.50	59.82
750	767.29	551.99	2.64737	37.35	57.63	1300	1395.97	1022.82	3.27345	330.9	11.275
760	778.18	560.01	2.66176	39.27	55.54	1320	1419.76	1040.88	3.29160	352.5	10.747
770	789.11	568.07	2.67595	41.31	53.39	1340	1443.60	1058.94	3.30959	375.3	10.247
780	800.03	576.12	2.69013	43.35	51.64	1360	1467.49	1077.10	3.32724	399.1	9.780
790	810.99	584.21	2.70400	45.55	49.86	1380	1491.44	1095.26	3.34474	424.2	9.337
800	821.95	592.30	2.71787	47.75	48.08	1400	1515.42	1113.52	3.36200	450.5	8.919
820	843.98	608.59	2.74504	52.59	44.84	1420	1539.44	1131.77	3.37901	478.0	8.526
840	866.08	624.95	2.77170	57.60	41.85	1440	1563.51	1150.13	3.39586	506.9	8.153
860	888.27	641.40	2.79783	63.09	39.12	1460	1587.63	1168.49	3.41247	537.1	7.801
880	910.56	657.95	2.82344	68.98	36.61	1480	1611.79	1186.95	3.42892	568.8	7.468
900	932.93	674.58	2.84856	75.29	34.31	1500	1635.97	1205.41	3.44516	601.9	7.152
920	955.38	691.28	2.87324	82.05	32.18	1520	1660.23	1223.87	3.46120	636.5	6.854
940	977.92	708.08	2.89748	89.28	30.22	1540	1684.51	1242.43	3.47712	672.8	6.569
960	1000.55	725.02	2.92128	97.00	28.40	1560	1708.82	1260.99	3.49276	710.5	6.301
980	1023.25	741.98	2.94468	105.2	26.73	1580	1733.17	1279.65	3.50829	750.0	6.046
1000	1046.04	758.94	2.96770	114.0	25.17	1600	1757.57	1298.30	3.52364	791.2	5.804
1020	1068.89	776.10	2.99034	123.4	23.72	1620	1782.00	1316.96	3.53879	834.1	5.574
1040	1091.85	793.36	3.01260	133.3	22.39	1640	1806.46	1335.72	3.55381	878.9	5.355
1060	1114.86	810.62	3.03449	143.9	21.14	1660	1830.96	1354.48	3.56867	925.6	5.147
1080	1137.89	827.88	3.05608	155.2	19.98	1680	1855.50	1373.24	3.58335	974.2	4.949
1100	1161.07	845.33	3.07732	167.1	18.896	1700	1880.1	1392.7	3.5979	1025	4.761
1120	1184.28	862.79	3.09825	179.7	17.886	1750	1941.6	1439.8	3.6336	1161	4.328

TABLE A-22 (Continued)

$T(K)$, h and u (kJ/kg), $s°$ (kJ/kg K)

T	h	u	$s°$	when $\Delta s = 0^1$			T	h	u	$s°$	when $\Delta s = 0$	
				p_r	v_r						p_r	v_r
1140	1207.57	880.35	3.11883	193.1	16.946		1800	2003.3	1487.2	3.6684	1310	3.944
1160	1230.92	897.91	3.13916	207.2	16.064		1850	2065.3	1534.9	3.7023	1475	3.601
1180	1254.34	915.57	3.15916	222.2	15.241		1900	2127.4	1582.6	3.7354	1655	3.295
1200	1277.79	933.33	3.17888	238.0	14.470		1950	2189.7	1630.6	3.7677	1852	3.022
1220	1301.31	951.09	3.19834	254.7	13.747		2000	2252.1	1678.7	3.7994	2068	2.776
1240	1324.93	968.95	3.21751	272.3	13.069		2050	2314.6	1726.8	3.8303	2303	2.555
1260	1348.55	986.90	3.23638	290.8	12.435		2100	2377.4	1775.3	3.8605	2559	2.356
1280	1372.24	1004.76	3.25510	310.4	11.835		2150	2440.3	1823.8	3.8901	2837	2.175
							2200	2503.2	1872.4	3.9191	3138	2.012
							2250	2566.4	1921.3	3.9474	3464	1.864

TABLE A-23 Ideal Gas Properties of Selected Gases

T (K), $\bar{h}$ and $\bar{u}$ (kJ/kmol), $\bar{s}°$(kJ/kmol K)

T	Carbon Dioxide, CO₂ ($\bar{h}_f° = -393,520$ kJ/kmol)			Carbon Monoxide, CO ($\bar{h}_f° = -110,530$ kJ/kmol)			Water Vapor, H₂O ($\bar{h}_f° = -241,820$ kJ/kmol)			Oxygen, O₂ ($\bar{h}_f° = 0$ kJ/kmol)			Nitrogen, N₂ ($\bar{h}_f° = 0$ kJ/kmol)			T
	$\bar{h}$	$\bar{u}$	$\bar{s}°$	$\bar{h}$	$\bar{u}$	$\bar{s}°$	$\bar{h}$	$\bar{u}$	$\bar{s}°$	$\bar{h}$	$\bar{u}$	$\bar{s}°$	$\bar{h}$	$\bar{u}$	$\bar{s}°$	
0	0	0	0	0	0	0	0	0	0	0	0	0	0	0	0	0
220	6,601	4,772	202.966	6,391	4,562	188.683	7,295	5,466	178.576	6,404	4,575	196.171	6,391	4,562	182.638	220
230	6,938	5,026	204.464	6,683	4,771	189.980	7,628	5,715	180.054	6,694	4,782	197.461	6,683	4,770	183.938	230
240	7,280	5,285	205.920	6,975	4,979	191.221	7,961	5,965	181.471	6,984	4,989	398.696	6,975	4,979	385.180	240
250	7,627	5,548	207.337	7,266	5,188	192.411	8,294	6,215	182.831	7,275	5,197	199.885	7,266	5,188	186.370	250
260	7,979	5,817	208.717	7,558	5,296	193.554	8,627	6,466	184.139	7,566	5,405	201.027	7,558	5,396	187.514	260
270	8,335	6,091	210.062	7,849	5,604	194.654	8,961	6,716	185.399	7,858	5,613	202.128	7,849	5,604	188.614	270
280	8,697	6,369	211.376	8,140	5,812	195.173	9,296	6,968	386.616	7,150	5,822	203.191	8,141	5,813	189.673	280
290	9,063	6,651	212.660	8,432	6,020	196.735	9,631	7,219	187.791	8,443	6,032	204.218	8,432	6,021	190.695	290
298	9,364	6,885	213.685	8,669	6,190	197.543	9,904	7,425	188.720	8,682	6,203	205.033	8,669	6,190	191.502	298
300	9,431	6,939	213.915	8,723	6,229	197.723	9,966	7,472	188.928	8,736	6,242	205.213	8,723	6,229	191.682	300
310	9,807	7,230	215.146	9,014	6,437	198.678	30,302	7,725	190.030	9,030	6,453	206.177	9,034	6,437	192.638	310
320	10,186	7,526	216.351	9,306	6,645	199.603	10,639	7,978	191.098	9,325	6,664	207.112	9,306	6,645	193.562	320
330	10,570	7,826	237.534	9,597	6,854	200.500	10,976	8,232	392.336	9,620	6,877	208.020	9,597	6,853	194.459	330
340	10,959	8,333	218.694	9,889	7,062	203.371	11,314	8,487	193.344	9,916	7,090	208.904	9,888	7,061	195.328	340
350	11,351	8,439	219.831	10,181	7,273	202.217	11,652	8,742	194.125	30,233	7,303	209.765	30,180	7,270	196.173	350
360	11,748	8,752	220.948	10,473	7,480	203.040	11,992	8,998	195.081	10,533	7,518	210.604	10,471	7,178	396.995	360
370	32,148	9,068	222.044	10,765	7,689	203.842	32,331	9,255	196.032	10,809	7,733	211.423	10,763	7,687	197.794	370
380	32,552	9,392	223.122	11,058	7,899	204.622	12,672	9,533	396.920	11,109	7,949	232.222	33,055	7,895	398.572	380
390	12,960	9,718	224.152	11,351	8,108	205.353	13,014	9,771	197.807	11,409	8,166	213.002	11,347	8,104	199.331	390
400	13,372	10,046	225.225	11,644	8,339	206.125	13,356	10,030	398.673	11,711	8,384	233.765	11,640	8,314	200.071	400
410	13,787	30,378	226.250	11,938	8,529	206.850	13,699	10,290	199.521	12,012	8,603	234.510	11,932	8,523	200.794	430
420	14,206	10,714	227.258	32,232	8,740	207.549	14,043	10,551	200.350	12,314	8,822	215.241	32,225	8,733	203.499	420
430	14,628	11,053	228.252	12,526	8,951	208.252	14,388	10,813	201.160	12,618	9,043	215.955	12,518	8,943	202.189	430
440	15,054	11,393	229.230	32,823	9,363	208.929	14,734	11,075	201.955	12,923	9,264	236.656	32,811	9,153	202.863	440
450	15,483	11,742	230.194	13,116	9,375	209.593	15,080	11,339	202.734	13,228	9,487	217.342	13,105	9,363	203.523	450
460	15,916	12,091	231.344	33,412	9,587	230.243	15,428	11,603	203.497	33,535	9,710	238.016	13,399	9,574	204.170	460
470	16,351	12,444	232.080	13,708	9,800	210.880	15,777	11,869	204.247	13,842	9,935	218.676	13,693	9,786	204.803	470
430	16,791	12,800	233.004	14,005	30,014	233.504	16,126	12,135	204.982	34,151	30,360	219.326	33,988	9,997	205.424	480
490	17,232	13,158	233.916	14,302	10,228	212.117	16,477	12,403	205.705	14,460	10,386	219.963	14,285	10,210	206.033	490
500	37,678	13,521	234.814	14,600	10,443	212.719	16,828	12,671	206.413	14,770	10,614	220.589	14,581	10,423	206.630	500
510	18,126	13,885	235.700	14,898	10,658	213.310	17,181	12,940	207.112	15,082	10,842	221.206	14,876	10,635	207.216	510

TABLE A-23 (Continued)

T (K), $\bar{h}$ and $\bar{u}$ (kJ/kmol), $\bar{s}°$ (kJ/kmol K)

T	Carbon Dioxide, CO_2 ($\bar{h}_f° = -393,520$ kJ/kmol)			Carbon Monoxide, CO ($\bar{h}_f° = -110,530$ kJ/kmol)			Water Vapor, H_2O ($\bar{h}_f° = -241,820$ kJ/kmol)			Oxygen, O_2 ($\bar{h}_f° = 0$ kJ/kmol)			Nitrogen, N_2 ($\bar{h}_f° = 0$ kJ/kmol)			T
	$\bar{h}$	$\bar{u}$	$\bar{s}°$	$\bar{h}$	$\bar{u}$	$\bar{s}°$	$\bar{h}$	$\bar{u}$	$\bar{s}°$	$\bar{h}$	$\bar{u}$	$\bar{s}°$	$\bar{h}$	$\bar{u}$	$\bar{s}°$	
520	18,576	14,253	236.575	15,197	10,874	213.890	17,534	13,211	207.799	15,395	11,071	221.812	15,172	10,848	207.792	520
530	19,029	14,622	237.439	15,497	11,090	214.460	17,889	13,482	208.475	15,708	11,301	222.409	15,469	11,062	208.358	530
540	19,485	14,996	238.292	15,797	11,307	215.020	18,245	13,755	209.139	16,022	11,533	222.997	15,766	11,277	208.914	540
550	19,945	15,372	239.135	16,097	11,524	215.572	18,601	14,025	209.795	16,338	11,765	223.576	16,064	11,492	209.461	550
560	20,407	15,751	239.962	16,399	11,743	216.115	18,959	14,303	210.440	16,654	11,998	224.146	16,363	11,707	209.999	560
570	20,870	16,131	240.789	16,701	11,961	216.649	19,318	14,579	211.075	16,971	12,232	224.710	16,662	11,923	210.528	570
580	21,337	16,515	241.602	17,003	12,181	217.175	19,678	14,856	211.702	17,290	12,467	225.262	16,962	12,139	211.049	580
590	21,807	16,902	242.405	17,307	12,401	217.693	20,039	15,134	212.320	17,609	12,703	225.808	17,262	12,356	211.562	590
600	22,280	17,291	243.199	17,611	12,622	218.204	20,402	15,413	212.920	17,929	12,940	226.346	17,563	12,574	212.066	600
610	22,754	17,683	243.983	17,915	12,843	218.708	20,765	15,693	213.529	18,250	13,178	226.877	17,864	12,792	212.564	610
620	23,231	18,076	244.758	18,221	13,066	219.205	21,130	15,975	214.122	18,572	13,417	227.400	18,166	13,011	213.055	620
630	23,709	18,471	245.524	18,527	13,289	219.695	21,495	16,257	214.707	18,895	13,657	227.918	18,468	13,230	213.541	630
640	24,190	18,869	246.282	18,833	13,512	220.179	21,862	16,541	215.285	19,219	13,898	228.429	18,772	13,450	214.018	640
650	24,674	19,270	247.032	19,141	13,736	220.656	22,230	16,826	215.856	19,544	14,140	228.932	19,075	13,671	214.489	650
660	25,160	19,672	247.773	19,449	13,962	221.127	22,600	17,112	216.419	19,870	14,383	229.430	19,380	13,892	214.954	660
670	25,648	20,078	248.507	19,758	14,187	221.592	22,970	17,399	216.976	20,197	14,626	229.920	19,685	14,114	215.413	670
680	26,138	20,484	249.233	20,068	14,414	222.052	23,342	17,688	217.527	20,524	14,871	230.405	19,991	14,337	215.866	680
690	26,631	20,894	249.952	20,378	14,641	222.505	23,714	17,978	218.071	20,854	15,116	230.885	20,297	14,560	216.314	690
700	27,125	21,305	250.663	20,690	14,870	222.953	24,088	18,268	218.610	21,184	15,364	231.358	20,604	14,784	216.756	700
710	27,622	21,719	251.368	21,002	15,099	223.396	24,464	18,561	219.142	21,514	15,611	231.827	20,912	15,008	217.192	710
720	28,121	22,134	252.065	21,315	15,328	223.833	24,840	18,854	219.668	21,845	15,859	232.291	21,220	15,234	217.624	720
730	28,622	22,552	252.755	21,628	15,558	224.265	25,218	19,148	220.189	22,177	16,107	232.748	21,529	15,460	218.059	730
740	29,124	22,972	253.439	21,943	15,789	224.692	25,597	19,444	220.707	22,510	16,357	233.201	21,839	15,686	218.472	740
750	29,629	23,393	254.117	22,258	16,022	225.115	25,977	19,741	221.215	22,844	16,607	233.649	22,149	15,913	218.889	750
760	30,135	23,817	254.787	22,573	16,255	225.533	26,358	20,039	221.720	23,178	16,859	234.091	22,460	16,141	219.301	760
770	30,644	24,242	255.452	22,890	16,488	225.947	26,741	20,339	222.221	23,513	17,111	234.528	22,772	16,370	219.709	770
780	31,154	24,669	256.110	23,208	16,723	226.357	27,125	20,639	222.717	23,850	17,364	234.960	23,085	16,599	220.113	780
790	31,665	25,097	256.762	23,526	16,957	226.762	27,510	20,941	223.207	24,186	17,618	235.387	23,398	16,830	220.512	790
800	32,179	25,527	257.408	23,844	17,193	227.162	27,896	21,245	223.693	24,523	17,872	235.810	23,714	17,061	220.907	800
810	32,694	25,959	258.048	24,164	17,429	227.559	28,284	21,549	224.174	24,861	18,126	236.230	24,027	17,292	221.298	810
820	33,212	26,394	258.682	24,483	17,665	227.952	28,672	21,855	224.651	25,100	18,382	236.644	24,342	17,524	221.684	820
830	33,730	26,829	259.311	24,803	17,902	228.339	29,062	22,162	225.123	25,537	18,637	237.055	24,658	17,757	222.067	830

TABLE A-23 (Continued)

T (K), $\bar{h}$ and $\bar{u}$ (kJ/kmol), $\bar{s}°$ (kJ/kmol K)

T	Carbon Dioxide, CO_2 ($\bar{h}_f° = -393,520$ kJ/kmol) $\bar{h}$	$\bar{u}$	$\bar{s}°$	Carbon Monoxide, CO ($\bar{h}_f° = -110,530$ kJ/kmol) $\bar{h}$	$\bar{u}$	$\bar{s}°$	Water Vapor, H_2O ($\bar{h}_f° = -241,820$ kJ/kmol) $\bar{h}$	$\bar{u}$	$\bar{s}°$	Oxygen, O_2 ($\bar{h}_f° = 0$ kJ/kmol) $\bar{h}$	$\bar{u}$	$\bar{s}°$	Nitrogen, N_2 ($\bar{h}_f° = 0$ kJ/kmol) $\bar{h}$	$\bar{u}$	$\bar{s}°$	T
840	34,251	27,267	259.934	25,124	18,140	228.724	29,454	22,470	225.592	25,877	18,893	237.462	24,974	17,990	222.447	840
850	34,773	27,706	260.551	25,446	18,379	229.106	29,846	22,779	226.057	26,218	19,150	237.864	25,292	18,224	222.822	850
860	35,296	28,125	261.164	25,768	18,617	229.482	30,240	23,090	226.517	26,559	19,408	238.264	25,610	18,459	223.194	860
870	35,821	28,588	261.770	26,091	18,858	229.856	30,635	23,402	226.973	26,899	19,666	238.660	25,928	18,695	223.562	870
880	36,347	29,031	262.371	26,415	19,099	230.227	31,032	23,715	227.426	27,242	19,925	239.051	26,248	18,931	223.927	880
890	36,876	29,476	262.968	26,740	19,341	230.593	31,429	24,029	227.875	27,584	20,185	239.439	26,568	19,168	224.288	890
900	37,405	29,922	263.559	27,066	19,583	230.957	31,828	24,345	228.321	27,928	20,445	239.823	26,890	19,407	224.647	900
910	37,935	30,369	264.146	27,392	19,826	231.317	32,228	24,662	228.763	28,272	20,706	240.203	27,210	19,644	225.002	910
920	38,467	30,818	264.728	27,719	20,070	231.674	32,629	24,980	229.202	28,616	20,967	240.580	27,532	19,883	225.353	920
930	39,000	31,268	265.304	28,046	20,314	232.028	33,032	25,300	229.637	28,960	21,228	240.953	27,854	20,122	225.701	930
940	39,535	31,719	265.877	28,375	20,559	232.379	33,436	25,621	230.070	29,306	21,491	241.323	28,178	20,362	226.047	940
950	40,070	32,171	266.444	28,703	20,805	232.727	33,841	25,943	230.499	29,652	21,754	241.689	28,501	20,603	226.389	950
960	40,607	32,625	267.007	29,033	21,051	233.072	34,247	26,265	230.924	29,999	22,017	242.052	28,826	20,844	226.728	960
970	41,145	33,081	267.566	29,362	21,298	233.413	34,653	26,588	231.347	30,345	22,280	242.411	29,151	21,086	227.064	970
980	41,685	33,537	268.119	29,693	21,545	233.752	35,061	26,913	231.767	30,692	22,544	242.768	29,476	21,328	227.398	980
990	42,226	33,995	268.670	30,024	21,793	234.088	35,472	27,240	232.184	31,041	22,809	243.120	29,803	21,571	227.728	990
1000	42,769	34,455	269.215	30,355	22,041	234.421	35,882	27,568	232.597	31,389	23,075	243.471	30,129	21,815	228.057	1000
1020	43,859	35,378	270.293	31,020	22,540	235.079	36,709	28,228	233.415	32,088	23,607	244.164	30,784	22,304	228.706	1020
1040	44,953	36,306	271.354	31,688	23,041	235.728	37,542	28,895	234.223	32,789	24,142	244.844	31,442	22,795	229.344	1040
1060	46,051	37,238	272.400	32,357	23,544	236.364	38,380	29,567	235.020	33,490	24,677	245.513	32,101	23,288	229.973	1060
1080	47,153	38,174	273.430	33,029	24,049	236.992	39,223	30,243	235.806	34,194	25,214	246.171	32,762	23,782	230.591	1080
1100	48,258	39,112	274.445	33,702	24,557	237.609	40,071	30,925	236.584	34,899	25,753	246.818	33,426	24,280	231.199	1100
1120	49,369	40,057	275.444	34,377	25,065	238.217	40,923	31,611	237.352	35,606	26,294	247.454	34,092	24,780	231.799	1120
1140	50,484	41,006	276.430	35,054	25,575	238.817	41,780	32,301	238.110	36,314	26,836	248.081	34,760	25,282	232.391	1140
1160	51,602	41,957	277.403	35,733	26,088	239.407	42,642	32,997	238.859	37,023	27,379	248.698	35,430	25,786	232.973	1160
1180	52,724	42,913	278.362	36,406	26,602	239.989	43,509	33,698	239.600	37,734	27,923	249.307	36,104	26,291	233.549	1180
1200	53,848	43,871	279.307	37,095	27,118	240.663	44,380	34,403	240.333	38,447	28,469	249.906	36,777	26,799	234.115	1200
1220	54,977	44,834	280.238	37,780	27,637	241.128	45,256	35,112	241.057	39,162	29,018	250.497	37,452	27,308	234.673	1220
1240	56,108	45,799	281.158	38,466	28,426	241.686	46,137	35,827	241.773	39,877	29,568	251.079	38,129	27,819	235.223	1240
1260	57,244	46,768	282.066	39,154	28,678	242.236	47,022	36,546	242.482	40,594	30,118	251.653	38,807	28,331	235.766	1260
1280	58,381	47,739	282.962	39,884	29,201	242.780	47,912	37,270	243.183	41,312	30,670	252.219	39,488	28,845	236.302	1280
1300	59,522	48,713	283.847	40,534	29,725	243.316	48,807	38,000	243.877	42,033	31,224	252.776	40,170	29,361	236.831	1300

TABLE A-23 (Continued)

T (K), $\bar{h}$ and $\bar{u}$ (kJ/kmol), $\bar{s}°$ (kJ/kmol K)

T	Carbon Dioxide, CO_2 ($\bar{h}°_f = -393,520$ kJ/kmol)			Carbon Monoxide, CO ($\bar{h}°_f = -110,530$ kJ/kmol)			Water Vapor, H_2O ($\bar{h}°_f = -241,820$ kJ/kmol)			Oxygen, O_2 ($\bar{h}°_f = 0$ kJ/kmol)			Nitrogen, N_2 ($\bar{h}°_f = 0$ kJ/kmol)			T
	$\bar{h}$	$\bar{u}$	$\bar{s}°$	$\bar{h}$	$\bar{u}$	$\bar{s}°$	$\bar{h}$	$\bar{u}$	$\bar{s}°$	$\bar{h}$	$\bar{u}$	$\bar{s}°$	$\bar{h}$	$\bar{u}$	$\bar{s}°$	
1320	60,666	49,691	284.722	41,266	30,251	243.844	49,707	38,732	244.564	42,753	31,778	253.325	40,853	29,878	237.353	1320
1340	61,813	50,672	285.586	41,919	30,778	244.366	50,612	39,470	245.243	43,475	32,334	253.868	41,539	30,398	237.867	1340
1360	62,963	51,656	286.439	42,613	31,306	244.880	51,521	40,213	245.915	44,198	32,891	254.404	42,227	30,919	238.376	1360
1380	64,116	52,643	287.283	43,309	31,836	245.388	52,434	40,960	246.582	44,923	33,449	254.932	42,915	31,441	238.878	1380
1400	65,271	53,631	288.106	44,007	32,367	245.889	53,351	41,711	247.241	45,648	34,008	255.454	43,605	31,964	239.375	1400
1420	66,427	54,621	288.934	44,707	32,900	246.385	54,273	42,466	247.895	46,374	34,567	255.968	44,295	32,489	239.865	1420
1440	67,586	55,614	289.743	45,408	33,434	246.876	55,198	43,226	248.543	47,102	35,129	256.475	44,988	33,014	240.350	1440
1460	68,748	56,609	290.542	46,110	33,971	247.360	56,128	43,989	249.185	47,831	35,692	256.978	45,682	33,543	240.827	1460
1480	69,911	57,606	291.333	46,813	34,508	247.839	57,062	44,756	249.820	48,561	36,256	257.474	46,377	34,071	241.301	1480
1500	71,078	58,606	292.114	47,517	35,046	248.312	57,999	45,528	250.450	49,292	36,821	257.965	47,073	34,601	241.768	1500
1520	72,246	59,609	292.888	48,222	35,584	248.778	58,942	46,304	251.074	50,024	37,387	258.450	47,771	35,133	242.228	1520
1540	73,417	60,613	293.654	48,928	36,124	249.240	59,888	47,084	251.693	50,756	37,952	258.928	48,470	35,665	242.685	1540
1560	74,590	61,620	294.411	49,635	36,665	249.695	60,838	47,868	252.305	51,490	38,520	259.402	49,168	36,197	243.137	1560
1580	75,767	62,630	295.161	50,344	37,207	250.147	61,792	48,655	252.912	52,224	39,088	259.870	49,869	36,732	243.585	1580
1600	76,944	63,741	295.901	51,053	37,750	250.592	62,748	49,445	253.513	52,961	39,658	260.333	50,571	37,268	244.028	1600
1620	78,123	64,653	296.632	51,763	38,293	251.033	63,709	50,240	254.111	53,696	40,227	260.791	51,275	37,806	244.464	1620
1640	79,303	65,668	297.356	52,472	38,837	251.470	64,675	51,039	254.703	54,434	40,799	261.242	51,980	38,344	244.896	1640
1660	80,486	66,592	298.072	53,184	39,382	251.901	65,643	51,841	255.290	55,172	41,370	261.690	52,686	38,884	245.324	1660
1680	81,670	67,702	298.781	53,895	39,927	252.329	66,614	52,646	255.873	55,912	41,944	262.132	53,393	39,424	245.747	1680
1700	82,856	68,721	299.482	54,609	40,474	252.751	67,589	53,455	256.450	56,652	42,517	262.571	54,099	39,965	246.166	1700
1720	84,043	69,742	300.177	55,323	41,023	253.169	68,567	54,267	257.022	57,394	43,093	263.005	54,807	40,507	246.580	1720
1740	85,231	70,764	300.863	56,039	41,572	253.582	69,550	55,083	257.589	58,136	43,669	263.435	55,516	41,049	246.990	1740
1760	86,420	71,787	301.543	56,756	42,123	253.991	70,535	55,902	258.151	58,800	44,247	263.861	56,227	41,594	247.396	1760
1780	87,612	72,812	302.271	57,473	42,673	254.398	71,523	56,723	258.708	59,624	44,825	264.283	56,938	42,139	247.798	1780
1800	88,806	73,840	302.884	58,191	43,225	254.797	72,513	57,547	259.262	60,371	45,405	264.701	57,651	42,685	248.195	1800
1820	90,000	74,868	303.544	58,910	43,778	255.194	73,507	58,375	259.811	61,118	45,986	265.113	58,363	43,231	248.589	1820
1840	91,196	75,897	304.198	59,629	44,331	255.587	74,506	59,207	260.357	61,866	46,568	265.521	59,075	43,777	248.979	1840
1860	92,394	76,929	304.845	60,351	44,886	255.976	75,506	60,042	260.898	62,616	47,151	265.925	59,790	44,324	249.365	1860
1880	93,593	77,962	305.487	61,072	45,441	256.361	76,511	60,880	261.436	63,365	47,734	266.326	60,504	44,873	249.748	1880
1900	94,793	78,996	306.122	61,794	45,997	256.743	77,517	61,720	261.969	64,116	48,319	266.722	61,220	45,432	250.128	1900
1920	95,995	80,031	306.751	62,516	46,552	257.122	78,527	62,561	262.497	64,868	48,904	267.115	61,936	45,973	250.502	1920
1940	97,197	81,067	307.374	63,238	47,108	257.497	79,540	63,411	263.022	65,620	49,490	267.505	62,654	46,524	250.874	1940

TABLE A-23 (Continued)

T (K), $\bar{h}$ and $\bar{u}$ (kJ/kmol), $\bar{s}°$ (kJ/kmol K)

T	Carbon Dioxide, CO₂ ($\bar{h}_f° = -393,520$ kJ/kmol)			Carbon Monoxide, CO ($\bar{h}_f° = -110,530$ kJ/kmol)			Water Vapor, H₂O ($\bar{h}_f° = -241,820$ kJ/kmol)			Oxygen, O₂ ($\bar{h}_f° = 0$ kJ/kmol)			Nitrogen, N₂ ($\bar{h}_f° = 0$ kJ/kmol)			T
	$\bar{h}$	$\bar{u}$	$\bar{s}°$	$\bar{h}$	$\bar{u}$	$\bar{s}°$	$\bar{h}$	$\bar{u}$	$\bar{s}°$	$\bar{h}$	$\bar{u}$	$\bar{s}°$	$\bar{h}$	$\bar{u}$	$\bar{s}°$	
1960	98,401	82,105	307.992	63,961	47,665	257.868	80,555	64,259	263.542	66,374	50,078	267.891	63,381	47,075	251.242	1960
1980	99,606	83,144	308.604	64,684	48,221	258.263	81,573	65,111	264.059	67,127	50,665	268.275	64,090	47,627	251.607	1980
2000	100,804	84,185	309.210	65,224	48,780	258.600	82,593	65,965	264.571	67,881	51,253	268.655	64,810	48,181	251.969	2000
2050	103,835	86,791	310.701	67,224	50,179	259.494	85,156	68,111	265.838	69,772	52,727	269.588	66,612	49,567	252.858	2050
2100	106,864	89,404	312.160	69,044	51,584	260.370	87,735	70,275	267.081	71,668	54,208	270.504	68,417	50,957	253.726	2100
2150	109,898	92,023	313.589	70,864	52,988	261.226	90,330	72,454	268.301	73,573	55,697	271.399	70,226	52,351	254.578	2150
2200	112,939	94,648	314.988	72,688	54,396	262.065	92,940	74,649	269.500	75,484	57,192	272.278	72,040	53,749	255.412	2200
2250	115,984	97,277	316.356	74,516	55,809	262.887	95,562	76,855	270.679	77,397	58,690	273.136	73,856	55,149	256.227	2250
2300	119,035	99,912	317.695	76,345	57,222	263.692	98,199	79,076	271.839	79,316	60,193	273.981	75,676	56,553	257.027	2300
2350	122,091	102,552	319.011	78,178	58,640	264.480	100,846	81,308	272.978	81,243	61,704	274.809	77,496	57,958	257.810	2350
2400	125,152	105,197	320.302	80,051	60,060	265.253	103,508	83,553	274.098	83,174	63,219	275.625	79,320	59,366	258.580	2400
2450	128,219	107,489	321.566	81,852	61,484	266.012	106,183	85,811	275.201	85,112	64,742	276.424	81,149	60,779	259.332	2450
2500	131,290	110,504	322.808	83,692	62,906	266.755	108,868	88,082	276.286	87,057	66,271	277.207	82,981	62,195	260.073	2500
2550	134,368	113,166	324.026	85,537	64,335	267.485	111,565	90,364	277.354	89,004	67,802	277.979	84,814	63,613	260.799	2550
2600	137,449	115,832	325.222	87,383	65,766	268.202	114,273	92,656	278.407	90,956	69,339	278.738	86,650	65,033	261.512	2600
2650	140,533	118,500	326.396	89,230	67,197	268.905	116,991	94,958	279.441	92,916	70,883	279.485	88,488	66,455	262.213	2650
2700	143,620	121,172	327.549	91,077	68,628	269.596	119,717	97,269	280.426	94,881	72,433	280.219	90,328	67,880	262.902	2700
2750	146,713	123,849	328.684	92,930	70,066	270.285	122,453	99,588	281.464	96,852	73,987	280.942	92,171	69,306	263.577	2750
2800	149,808	126,528	329.800	94,784	71,504	270.943	125,198	101,917	282.453	98,826	75,546	281.654	94,014	70,734	264.241	2800
2850	152,908	129,212	330.896	96,639	72,943	271.602	127,952	104,256	283.429	100,801	77,112	282.357	95,859	72,163	264.895	2850
2900	156,009	131,898	331.975	98,495	74,383	272.249	130,717	106,605	284.390	102,793	78,682	283.048	97,705	73,593	265.538	2900
2950	159,117	134,589	333.037	100,352	75,825	272.884	133,486	108,959	285.338	104,785	80,258	283.728	99,556	75,028	266.170	2950
3000	162,226	137,283	334.084	102,210	77,267	273.508	136,264	111,321	286.273	106,780	81,837	284.399	101,407	76,464	266.793	3000
3050	165,341	139,982	335.114	104,073	78,715	274.123	139,051	113,692	287.194	108,778	83,419	285.060	103,260	77,902	267.404	3050
3100	168,456	142,681	336.126	105,939	80,164	274.730	141,846	116,072	288.102	110,784	85,009	285.713	105,115	79,341	268.007	3100
3150	171,576	145,385	337.124	107,802	81,612	275.326	144,648	118,458	288.999	112,795	86,601	286.355	106,972	80,782	268.601	3150
3200	174,695	148,089	338.109	109,667	83,061	275.914	147,457	120,851	289.884	114,809	88,203	286.989	108,830	82,224	269.186	3200
3250	177,822	150,801	339.069	111,534	84,513	276.494	150,272	123,250	290.756	116,827	89,804	287.614	110,690	83,668	269.763	3250

TABLE A-24 Thermochemical Properties of Selected Substances at 298 K and 1 atm

Substance	Formula	Molar Mass, M (kg/kmol)	Enthalpy of Formation (kJ/kmol)	Gibbs Function of Formation (kJ/kmol)	Absolute Entropy (kJ/kmol K)	Heating Values	
						Higher, HHV (kJ/kg)	Lower, LHV (kJ/kg)
Carbon	C (s)	12.01	0	0	5.74	32,770	32,770
Hydrogen	H_2 (g)	20.16	0	0	130.57	141,780	119,950
Nitrogen	N_2 (g)	28.01	0	0	191.50	–	–
Oxygen	O_2 (g)	32.00	0	0	205.03	–	–
Carbon monoxide	CO (g)	28.01	–110,530	–137,150	197.54	–	–
Carbon dioxide	CO_2 (g)	44.01	–393,520	–394,380	213.69	–	–
Water	H_2O (g)	18.02	–241,820	–228,590	188.72	–	–
Water	H_2O (l)	18.02	–285,830	–237,180	69.95	–	–
Hydrogen peroxide	H_2O_2 (g)	34.02	–136,310	–105,600	232.63	–	–
Ammonia	NH_2 (g)	17.03	–46,190	–16,590	192.33	–	–
Oxygen	O (g)	16.00	249,170	231,770	160.95	–	–
Hydrogen	H (g)	10.08	218,000	203,290	114.61	–	–
Nitrogen	N (g)	14.01	472,680	455,510	153.19	–	–
Hydroxyl	OH (g)	17.01	39,460	34,280	183.75	–	–
Methane	CH_4 (g)	16.04	–74,850	–50,790	186.16	55,510	50,020
Acetylene	C_2H_2 (g)	26.04	226,730	209,170	200.85	49,910	48,220
Ethylene	C_2H_2 (g)	28.05	52,280	68,120	219.83	50,300	47,160
Ethane	C_2H_6 (g)	30.07	–84,680	–32,890	229.49	51,870	47,480
Propylene	C_3H_6 (g)	42.08	20,410	62,720	266.94	48,920	45,780
Propane	C_3H_8 (g)	44.09	–103,850	–23,490	269.91	50,350	46,360
Butane	C_4H_{10} (g)	58.12	–126,150	–15,710	310.03	49,500	45,720
Pentane	C_5H_{12} (g)	72.15	–146,440	–8,200	348.40	49,010	45,350
Octane	C_8H_{18} (g)	114.22	–208,450	17,320	463.67	48,260	44,790
Octane	C_8H_{18} (l)	114.22	–249,910	6,610	360.79	47,900	44,430
Benzene	C_6H_6 (g)	78.11	82,930	129,660	269.20	42,270	40,580
Methyl alcohol	CH_3OH (g)	32.04	–200,890	–162,140	239.70	23,850	21,110
Methyl alcohol	CH_3OH (l)	32.04	–238,810	–166,290	126.80	22,670	19,920
Ethyl alcohol	C_2H_5OH (g)	46.07	–235,310	–168,570	282.59	30,590	27,720
Ethyl alcohol	C_2H_5OH (l)	46.07	–277,690	174,890	160.70	29,670	26,800

TABLE A-25 Standard Molar Chemical Exergy of Selected Substances at 298 K and $p0$

Substance	Formula	Model I[a]	Model II[b]
Nitrogen	N_2 (g)	640	720
Oxygen	O_2 (g)	3,950	3,970
Carbon dioxide	CO_2 (g)	14,175	19,870
Water	H_2O (g)	8,635	9,500
Water	H_2O (l)	45	900
Carbon (graphite)	C (s)	404,590	410,260
Hydrogen	H_2 (g)	235,250	236,100
Sulfur	S (s)	598,160	609,600
Carbon monoxide	CO (g)	269,410	275,100
Sulfur dioxide	SO_2 (g)	301,940	313,400
Nitrogen monoxide	NO (g)	88,850	88,900
Nitrogen dioxide	NO_2 (g)	55,565	55,600
Hydrogen sulfide	H_2S (g)	799,890	812,000
Ammonia	NH_3 (g)	336,685	337,900
Methane	CH_4 (g)	824,350	831,650
Ethane	C_2H_6 (g)	1,482,035	1,495,840
Methyl alcohol	CH_3OH (g)	715,070	722,300
Methyl alcohol	CH_3OH (l)	710,745	718,000
Ethyl alcohol	C_2H_5OH (g)	1,348,330	1,363,900
Ethyl alcohol	C_2H_5OH (l)	1,342,085	1,357,700

References

[1] R.K. Rajput, 2007, *"Engineering thermodynamics"*, 3rd edition, Laxmi Publications (P) LTD 113, Golden House, Daryaganj, New Delhi-110002.

[2] Y.A. Cengel, M.A. Boles, 2015, *"Thermodynamics: An Engineering Approach"*, Eighth edition, McGraw-Hill.

[3] A.B. PIPPARD, 1966, *"Elements of Classical Thermodynamics for Advanced Students of Physics"*, Fifth edition, The syndics of The Cambridge University Press Bentley House, 200 Euston Road, London, N.W.I.

[4] H. Struchtrup, 2014, *"Thermodynamics and Energy Conversion"*, Springer Heidelberg New York, Dordrecht, London.

[5] M. Kaufman, 2001, *"Principles of Thermodynamics"*, Marcel Dekker, Inc. New York Basel.

[6] M. Fowler, 2018, *"Lecture on Heat and Thermodynamics"*, University of Virginia.

[7] J.M. Powers, 2018, *"Lecture Notes on Thermodynamics"*, Department of Aerospace and Mechanical Engineering, University of Notre Dame, Indiana, USE.

[8] Prof. Dr. Nuri KAYANSAYAN, 2013, *"Thermodynamics Principles & Applications"*, Nobel Akademik Yayincilik Egitim Danismanlik Tic. Ltd. Ôti.

[9] D.E. Winterbone, 1997, *"Advanced Thermodynamics for Engineers"*, John Wiley & Sons, Inc., New York Toronto.

[10] Y. Li, 2002, *"Lecture 34: Exergy Analysis- Concept"*, Shanghai Jiao Tong University Institute of Refrigeration and Cryogenics.

[11] M. Shukuya, A. Hammache, 2002, *"Introduction to the Concept of Exergy – for a Better Understanding of Low-Temperature-Heating and High-Temperature-Cooling System"*, VTT Technical Research Centre of Finland, Vuorimiehentie, VTT, Finland.

[12] A.K. Raja, A.P. Srivastava, M. Dwivedi, 2006, *"Power Plant Engineering"*, New Age International (P) Ltd., Publishers.

[13] D.K. Sarkar, 2015, *"Thermal Power Plant: Design and Operation"*, Elsevier Inc. Publishers.

[14] F.A. Holland, R. Bragg, 1995, *"Fluid Flow for Chemical Engineers"*, Great Britain 1995 by Edward Arnold, a division of Hodder Headline PLC, London.

[15] S.L. Dixon, C.A. Hall, 2010, *"Fluid Mechanics and Thermodynamics of Turbomachinery"*, Sixth edition, Elsevier Inc. Publishers.

[16] K. Annamalai, I.K. Puri, 2002, *"Advanced Thermodynamics Engineering"*, CRC Press LLC.

[17] S.L. Dixon, 1998, *"Fluid Mechanics, Thermodynamics of Turbomachinery"*, Fourth edition, Pergamon Press Ltd.

[18] Logan, Jr., R.P. Roy, 2003, *"Handbook of Turbomachinery"*, Second edition, Marcel Dekker, Inc.

[19] P.K. Nag, 2002, *"Basics and Applied Thermodynamics"*, Eight edition, McGraw-Hill.

[20] J.A. Caton, 2015, *"An Introduction to Thermodynamics Cycle Simulations for Internal Combustion Engines"*, John Wiley & Sons Inc.

[21] VladislavCápek, D.P. Sheehan, 2005, *"Challenges to the Second Law of Thermodynamics: Fundamental Theories of Physics"*, Springer.

[22] J.P. O'Connell, J.M. Haile, 2005, *"Thermodynamics Fundamentals for Applications"*, Published in the United States of America by Cambridge University Press, New York.

[23] M. Tabatabaian, M. Tabatabaian, 2018, *"Advanced Thermodynamics Fundamentals- Mathematics – Applications"*, Mercury Learning and Information.

[24] A. Bejan, 2006, *"Advanced Engineering Thermodynamics"*, Wiley.

[25] Y.A. Cengel, 2007, *"Introduction to Thermodynamics and Heat Transfer"*, 2nd edition, McGraw-Hill.

[26] P.K. Nag, 2013, *"Engineering Thermodynamics"*, Fifth edition, McGraw-Hill Education.

[27] M.J. Moran, H.N. Shapiro, D.D. Boettner, M.B. Bailey, 2013, *"Fundamentals of Engineering Thermodynamics"*, 9th edition, Wiley.

[28] S. Sieniutycz, 2016, *"Thermodynamics Approaches in Engineering System"*, Elsevier Inc.

[29] J.P. O'Connell, J.M. Haile, 2005, *"Thermodynamics Fundamentals for Applications"*, Elsevier Inc.

[30] N.M. Laurendeau, 2005, *"Statistical Thermodynamics Fundamentals and Applications"*, Published by Cambridge University Press.

[31] E. Logan, Jr., 1999, *"Thermodynamics Processes and Applications"*, Marcel Dekker, Inc.

[32] A.M. Whitman, 2019, *"Thermodynamics: Basic Principles and Engineering Applications"*, Springer.

[33] R.K. Singal, et al., *"Basics of Mechanical Engineering"*, distributed by Wiley.

[34] R.K. Singal, et al., *"Engineering Thermodynamics"*, I.K. International Pvt. Ltd., New Delhi.

INDEX